E685   COMPLIMENTARY COPY   ✓

DATE   15.5.80   QUANTITY
          08-8
          EXPLORINGMETRIC DRAFTING
TITLE   Walker
AUTHOR $14.95E ✓
PRICE
CONTACT   Steve Walker

THIS BOOK IS PRESENTED AS A CLASS TEXT,
YOUR COMMENTS WOULD BE APPRECIATED.

GENERAL
PUBLISHING CO. LIMITED

30 LESMILL ROAD, DON MILLS, ONT.
M3B 2T6   TELE. (416) 445-3333

Exploring metric drafting

basic fundamentals

by

JOHN R. WALKER

Bel Air High School
Bel Air, Maryland

South Holland, Illinois
THE GOODHEART-WILLCOX COMPANY, INC.
Publishers

Library of Congress Cataloging in Publication Data

Walker, John R.
  Exploring metric drafting.

  Includes index.
  1. Mechanical drawing.   2. Metric system.   I. Title.
T353.W233   604'.2   79—24019
ISBN 0—87006—289—1

INTRODUCTION

EXPLORING METRIC DRAFTING is a first course which teaches Drafting Fundamentals and Basic Constructions. As you proceed, you will become familiar with methods and processes used by industry. You will make many of the Drafter's skills your own. A course in metric drafting will help you to develop the capacity to plan in an orderly fashion, to interpret the ideas of others, and to express in an understandable manner.

EXPLORING METRIC DRAFTING is the metric version of EXPLORING DRAFTING, also published by Goodheart-Willcox. This metricated text contains metric dimensioned drawings. It will be revised as new drafting standards are finalized and adopted.

The problems presented in this text are similar to the problems contained in the original EXPLORING DRAFTING. They have been metricated to take advantage of standardized metric stock sizes. They are not problems where the original inch dimensions were simply converted to metric units. Metric drafting is much more than that.

For the best learning situation, avoid converting the metric units to inches and fractions and making your measurements in these units. Use a metric scale. Do not even think in inches. Otherwise, confusion may result and mistakes will be made.

Drafting is the "Language of Industry." It offers many opportunities for a lifetime of challenging and personally rewarding careers. Will you accept that challenge?

John R. Walker

CONTENTS

Typical industry photo. Hundreds of drawings were required to plan and build the hotel
in which this 21.3 metre long truss was used.   (Bethlehem Steel Corp.)

Unit 1

WHY STUDY
METRIC BASE DRAFTING

DRAFTING is the part of industry concerned with the preparation of drawings needed to develop and manufacture modern products.

Drawings often are the best means available to explain or show our ideas, Fig. 1-1. Drafting frequently is called a "universal language" because it communicates ideas in graphic or picture form. Like other languages, symbols (lines and figures) that have special and specific meanings are used to accurately describe the shape, size, type of material, finish and fabrication of an object. The symbols have been standardized over most of the world, making it possible to interpret and understand drawings made in other countries. See Fig. 1-2.

Drafting also is the "language of industry." It is a precise language because drawings provide the craftworker with the information needed to produce the product the designer had in mind, Fig. 1-3. It involves

Fig. 1-1. Drawings are the best way to show how to construct something as complex as this helicopter. Can you imagine trying to explain how to make a helicopter using words only?   (Sikorsky Aircraft)

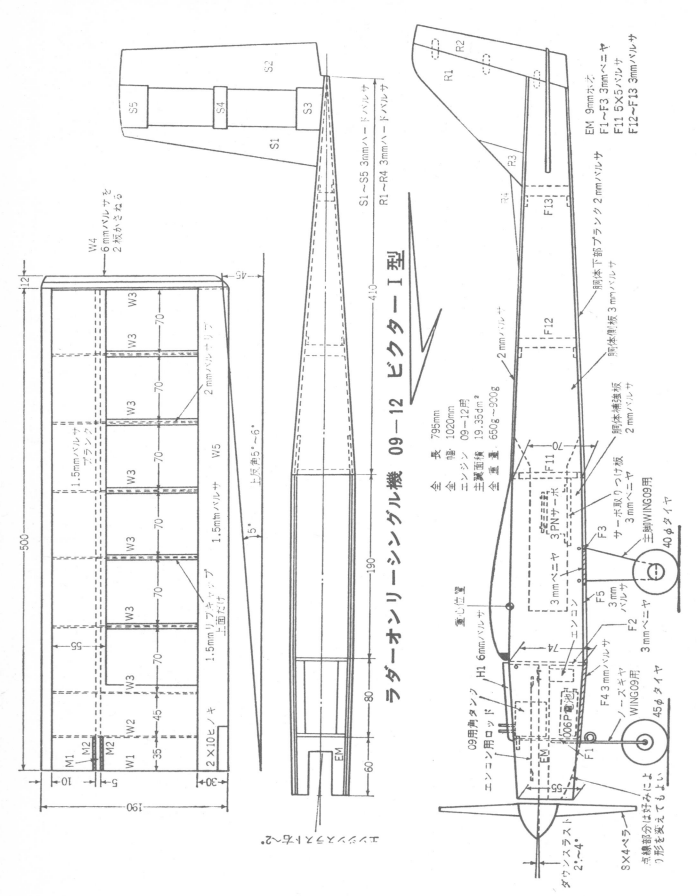

ラダーオンリーシングル機 09－12 ビクターⅠ型

全　長　795mm
全　幅　1020mm
エンジン　09～12用
主翼面積　19.35dm²
全重量　650g～900g

Fig. 1-2. Although the written instructions are in Japanese, it would be possible for you to use this plan to construct this radio controlled model airplane, if you understand drawings and the metric system.

Fig. 1-3. Drafting is the "language of industry." Drawings show craftworkers what the designer had in mind.   (American Motors)

the recording of necessary production information on paper, film or tape, Fig. 1-4.

Drafting is very important to our modern world. It would be difficult to name an occupation that does not require the ability to read and understand drawings.

It is quite obvious that the craftworkers who build our homes, make our cars or produce our TV sets must use drawings. YOU use drawings when you assemble a model, Fig. 1-5.

Many specialized fields of drafting have been developed: aerospace, architectural, automotive styl-

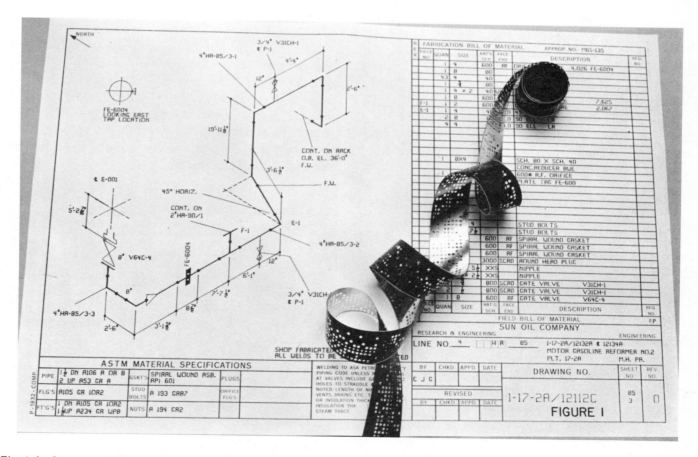

Fig. 1-4. Some specialized drafting can be done by machine. The required information is "punched" into a plastic tape. The tape is fed into a computer-like machine which interprets the hole sequence in the tape and prints it on a sheet of paper. This drawing of a piping diagram in an oil refinery was made in this manner.   (Sun Oil Co.)

9

Fig. 1-5. You use drawings when you construct a model.

ing, electronic and electrical drafting, structural drafting, engineering graphics, technical drawing and topographical drafting.

Basically, they use the same drafting equipment and employ similar drafting techniques. However, the type of work done varies greatly.

But why study metric base drafting? For that matter, what metric units of measure will you use and how will you use them?

To answer the first question, the United States is "going metric." Drafting will play a very important role in our conversion to the metric system of measure.

As to the metric units of measure you will use, the full conversion to metrics by the United States will take a very long time. No specific time schedule has been established. For the present, your knowing and understanding the following metric units will be a big step towards full understanding of the metric system. The chart in Fig. 1-6 will make the learning easier.

ISO, SI AND ANSI

In the process of learning metrics, you will find references to the terms ISO, SI and ANSI. These terms have the following meaning:

ISO is the International Organization for Standardization. This international specialized agency for standardization is composed of the national standards bodies of 80 countries. They are concerned with the terminology, dimensions and test methods (includes safety and quality) for international standards.

STANDARDS define and allow measurement of length, volume, weight, time and other values. They are independent of environmental conditions such as temperature and pressure.

SI stands for the Systeme International d'Unites and refers to the metric system in its most modern and perfected form. There is only one SI unit for each physical quantity. See Fig. 1-7. The system is simple. Multiples and submultiples of an SI unit are related to the unit by powers of 10. There is also a consistent set of prefixes for naming the multiples and submultiples of the unit.

TO MEASURE	METRIC UNIT USED	COMPARES TO THIS
Length	millimetre (mm) centimetre (cm) metre (m) kilometre (km)	Thickness of paper clip Width of paper clip About 39 1/2 inches About 6/10 mile
Area	hectare (ha)	About 2 1/2 acres
Weight	gram (g) kilogram (kg) metric ton (t)	Weight of paper clip About 2.2 pounds 2240 pounds
Volume	litre (L) millilitre (ml)	1 quart, 2 ounces 1/5 teaspoon
Temperature	degree Celsius (°C)	Water freezes at 0°C, boils at 100°C. To convert to Fahrenheit roughly: double °C and add 30

Fig. 1-6. How metric units roughly compare with English measure and familiar household items.

kilometre (km)	=	1 000 metres (thousands)
hectometre (hm)	=	100 metres (hundreds)
dekametre (dam)	=	10 metres (tens)
metre (m)	=	1 metre (unit of linear measure)
decimetre (dm)	=	0.1 metre (tenths)
centimetre (cm)	=	0.01 metre (hundredths)
millimetre (mm)	=	0.001 metre (thousandths)
micrometre (μ m)	=	0.000 001 metre (millionths)

Fig. 1-7. The most common linear units in the metric system.

ANSI is the American National Standards Institute. This is the organization that represents the United States in the International Organization for Standardization. It approves and distributes the ISO standards in the United States.

TEST YOUR KNOWLEDGE - UNIT 1

(Write answers on a separate sheet of paper.)
1. What does the term drafting mean?
2. Drawings are often used because:
   a. They are easy to make.
   b. They are the best means available to explain or show many ideas.
   c. People who cannot read can understand them.
   d. All of the above.
   e. None of the above.
3. Drafting is also called:
   a. A universal language.
   b. The language of industry.
   c. A picture language.
   d. All of the above.
   e. None of the above.
4. In addition to the craftworkers who make the things we use every day, list several other people who make use of drawings.
5. List five specialized fields of drafting.
6. With reference to the metric system, what is the meaning of each of the following terms:
   ISO _____ .
   SI _____ .
   ANSI _____ .
7. _____ is the unit of linear measure in the metric system.
8. What is the metric equivalent for the term Fahrenheit?

OUTSIDE ACTIVITIES

1. Secure samples of drawings used by the following industries:
   a. Aerospace.
   b. Architecture.
   c. Automotive styling.
   d. Electrical and electronic.
   e. Structural.
   f. Manufacturing (technical drawings).
   g. Map making.
2. Make a collection of pictures (magazine clippings, photographs, etc.) that show products made by industries listed in No. 1.
3. Obtain copies of drawings made in a foreign country.
4. Visit a local architect who designs residences. After discussing a project in work, prepare a report on the steps normally followed when designing a home for a client. Prepare your questions carefully before your visit.
5. Visit a local surveyor and make a report on this type of work. Borrow sample surveys for a bulletin board display.

The architectural drafter must have the technical knowledge required to carry out the ideas of the architect.   (Prestressed Concrete Institute)

The fabrication and erection of steel structures like the Lamoille River (Vt.) Bridge requires many drawings. With drawings the drafter conveys, in technical language, all the information required to fabricate the many structural members.   (Bethlehem Steel)

Unit 2
SHOP SKETCHING

SHOP SKETCHING is one of many drafting techniques. It is a convenient and rapid way of putting ideas into visual form. See Fig. 2-1.

A good sketch shows the shape of the object, and provides dimensions and special instructions on how the object is to be made and finished.

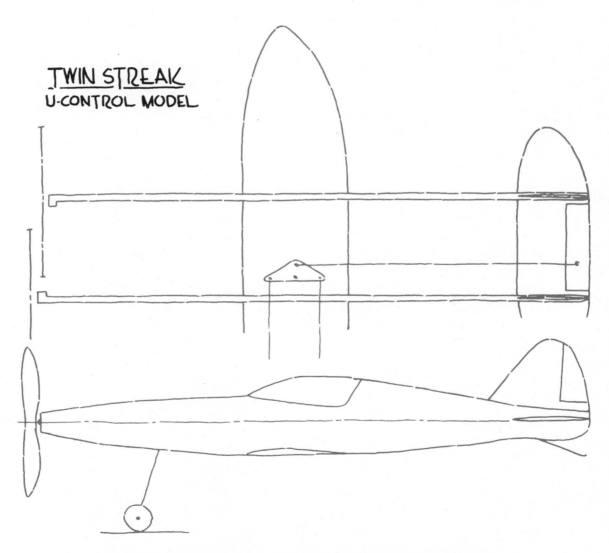

TWIN STREAK
U-CONTROL MODEL

Fig. 2-1. Sketch of a model plane design. Sketching is a convenient and rapid way of putting ideas into visual form.

13

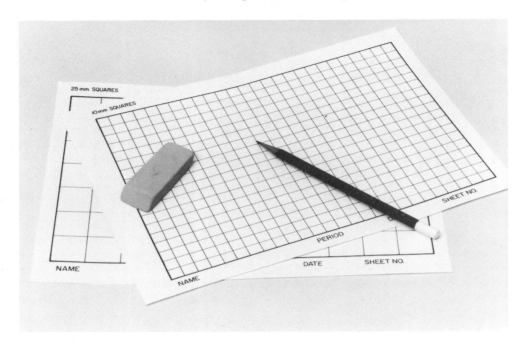

Fig. 2-2. Graph paper is useful in sketching. A good eraser is also a necessity.

Sketching does not require a great deal of equipment. Properly sharpened F, 2H or HB grade pencils and sheets of standard 8 1/2 x 11 in. (210 x 297 mm) letter size paper are well suited for this purpose. See Fig. 2-2. The paper can be plain or cross sectioned (graph or squares). A good eraser is also needed.

In sketching, a line is drawn by making a series of short strokes, Fig. 2-3. Draw light (thin) construction lines first—corrections are easier to make. Horizontal lines are drawn from left to right; vertical lines are drawn from the top down.

The instructions that follow tell and show how to sketch basic geometric shapes. You will find that even the most complex drawings are made up of a combination of these basic geometric shapes.

ALPHABET OF LINES (SKETCHED)

A drafter uses lines of various weights (thickness) to make a drawing. Each line has a special meaning, Fig. 2-4. Contrast between the various line weights or thicknesses helps to make a drawing easier to read. It is essential that you learn this ALPHABET OF LINES:

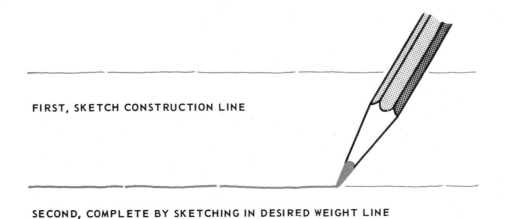

FIRST, SKETCH CONSTRUCTION LINE

SECOND, COMPLETE BY SKETCHING IN DESIRED WEIGHT LINE

Fig. 2-3. When sketching, a line is drawn by making a series of short strokes.

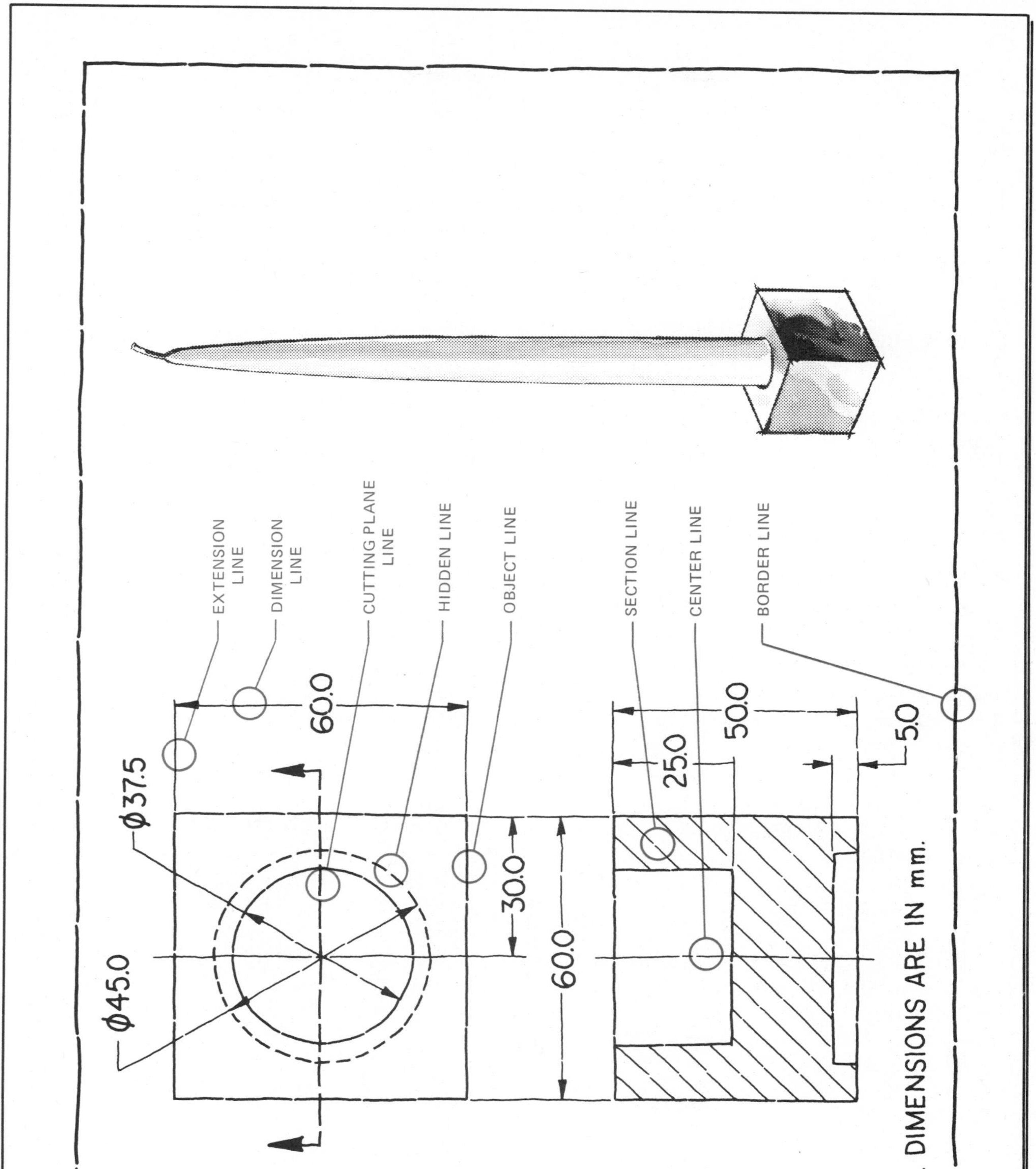

EXTENSION
LINE

DIMENSION
LINE

CUTTING PLANE
LINE

HIDDEN LINE

OBJECT LINE

SECTION LINE

CENTER LINE

BORDER LINE

60.0

Ø 37.5

Ø 45.0

30.0

60.0

50.0

25.0

5.0

ALL DIMENSIONS ARE IN mm.

Fig. 2-4. ALPHABET OF LINES as used in sketching a project.

CONSTRUCTION AND GUIDE LINE

CONSTRUCTION LINES are used to lay out drawings. GUIDE LINES are used when lettering to help you keep the lettering uniform in height. These lines are drawn lightly, using a pencil with the lead sharpened to a long conical point.

BORDER LINE

The BORDER LINE is the heaviest (thickest) line used in sketching. First, draw light construction lines as a guide, then go over them using a pencil with a heavy rounded point to provide the border lines.

OBJECT LINE

The OBJECT LINE is a heavy line, but slightly less in thickness than the border line. The object line indicates visible edges. In sketching object lines, use a pencil with a medium lead and a rounded point.

HIDDEN LINE

HIDDEN LINES are used to indicate or show hidden features of a part. The hidden line is made up of broken lines with dashes about 4.0 mm long with about 1.0 mm spaces. It is medium weight and less prominent than the visible line.

EXTENSION LINE

EXTENSION LINES are the same weight as dimension lines. These lines indicate points from which the dimensions are given. The extension line begins 1.5 mm away from the view and extends about 1.5 mm past the last dimension line.

75.0

DIMENSION LINE

DIMENSION LINES generally terminate in arrowheads (about 3.0 mm long) at the ends. Generally, they are placed between two extension lines. A break is made, usually in the center, to place the dimension. A dimension line is placed a minimum of 10 mm away from the drawing. It is a fine line and is drawn with a pencil sharpened to a long conical point.

CENTER LINE

CENTER LINES are made up of alternate long dashes (10 to 40 mm long) and short dashes and spaces (about 3.0 mm and 1.5 mm respectively). Center lines are drawn about the same weight as dimension and extension lines. They are used to locate centers of symmetrical objects.

CUTTING PLANE LINE

CUTTING PLANE LINES indicate where an object has been cut to show internal features. Two types are acceptable. However, a cutting plane line made of 6.5 mm dashes with 1.0 mm spacing is recommended. The second type uses a long dash (15-40 mm) and short 3.0 mm dashes with 1.0 mm spacing. A cutting plane line is slightly heavier than an object line and is drawn using a pencil with a rounded point.

SECTION LINE

SECTION LINES are used when drawing inside features of an object to indicate the surfaces exposed by the cutting plane line. Section lines are also used to indicate general classifications of materials. These lines, light in weight, are drawn with a pencil sharpened to a long conical point.

HOW TO SKETCH A HORIZONTAL LINE

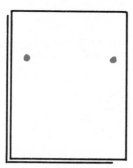

1. Mark off two points spaced a distance equal to the length of the line to be drawn. The points should be parallel to the top or bottom edge of your paper.

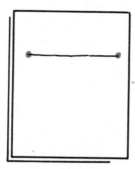

2. Move your pencil back and forth and connect these points with a construction line.

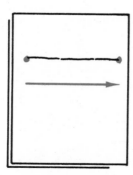

3. Start from the left point and sketch an object line to the right point. This line is sketched over the construction line.

HOW TO SKETCH A VERTICAL LINE

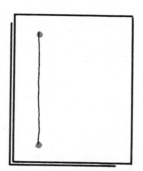

1. Mark off two points spaced a distance equal to the length of the line to be drawn. The two points should be parallel to the right or left edge of the sheet. Move your pencil back and forth and connect these points with a construction line.

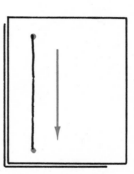

2. Start from the top point and sketch down and over the construction line to draw the desired line.

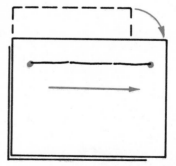

3. Vertical lines can also be sketched by rotating the paper into a horizontal position and proceeding as explained in HOW TO SKETCH A HORIZONTAL LINE.

HOW TO SKETCH AN INCLINED LINE

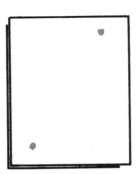

1. Mark off two points at the desired angle. Connect these points with a construction line.

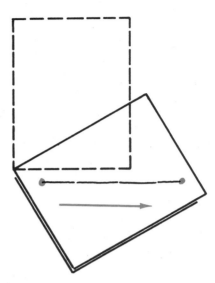

3. Inclined lines can also be sketched by rotating the sheet so the points are in a horizontal position. Sketch the line as previously described.

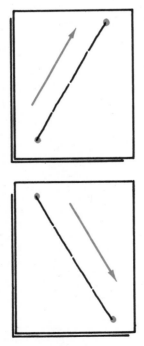

2. Sketch the desired weight line over the construction line. Sketch in the directions illustrated. Sketch up when the line inclines to the right. Sketch down when the line inclines to the left.

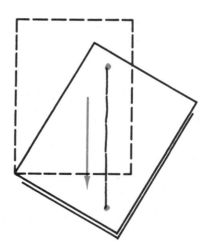

4. For some sketching problems, it may be easier to rotate the paper so the points are in a vertical position. Proceed as explained in HOW TO SKETCH A VERTICAL LINE.

HOW TO SKETCH SQUARES AND RECTANGLES

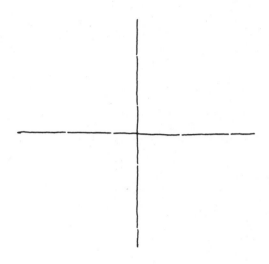

1. Sketch a horizontal line and a vertical line (axes).

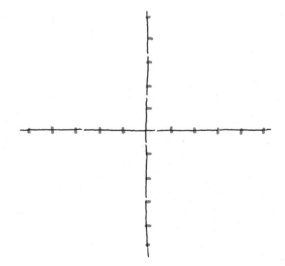

2. Begin at the intersection of these lines and lay out equal units on both lines in both directions. For example: If you want to draw a 50 mm square, you would estimate a unit of 5.0 mm and mark off five of these units on the vertical axis above and below the horizontal axis. Lay out the horizontal axis in the same manner.

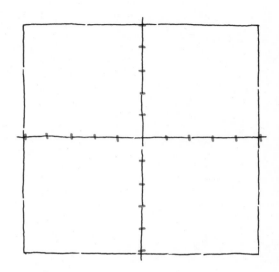

3. Sketch construction lines through the desired points.

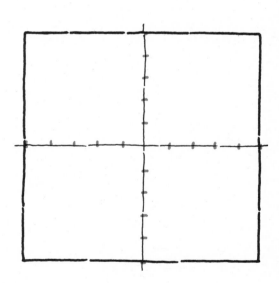

4. Go over the construction lines, forming the square to produce the desired weight line.

5. Rectangles are sketched in the same way except that you will have more units on one axis (line) than the other axis (line).

HOW TO SKETCH ANGLES

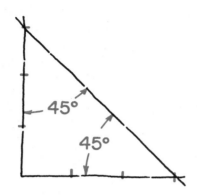

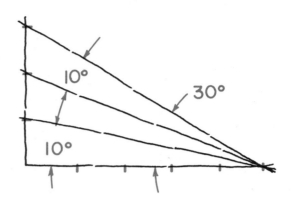

1. Sketch vertical and horizontal construction lines. These lines will form a 90 deg. or RIGHT ANGLE.

2. A 45 deg. angle is sketched by marking off an equal number of units on both lines. Connect the last unit of each line. This will form a 45 deg. angle with the vertical and the horizontal lines.

4. Other angles may be drawn by sketching an angle and subdividing this into the approximate number of degrees required. Example: Dividing a 30 deg. angle into thirds will give a 10 deg. angle.

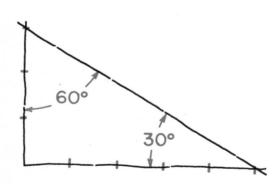

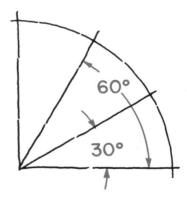

3. To sketch 30 and 60 deg. angles, mark off three units on one line and five units on the other line. Connecting the last unit on each line will give the required angles.

5. Another method used to develop angles in sketching, is to sketch a quarter circle and divide the resulting arc into the desired divisions. Example: Dividing the arc into three parts will give 30 and 60 deg. angles.

HOW TO SKETCH CIRCLES

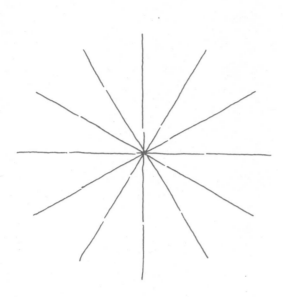

1. Sketch vertical, horizontal and inclined axes.

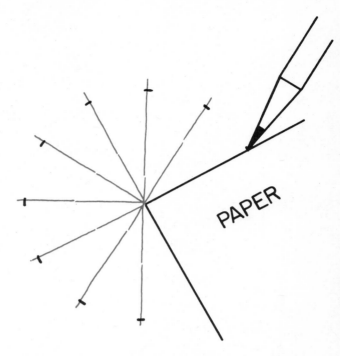

3. The radius units can be quickly and accurately located. Mark off the desired radius on a piece of paper, then use the paper as a measuring tool.

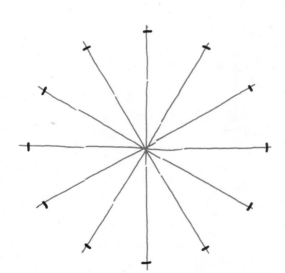

2. Mark off units equal to the radius of the required circle on each axis.

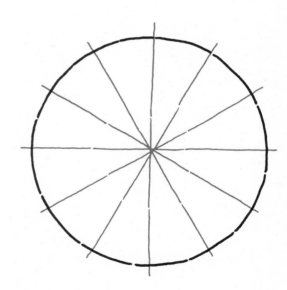

4. Sketch a construction line through the points. When satisfied with the construction line, fill it in with a line of the desired weight.

HOW TO SKETCH AN ARC

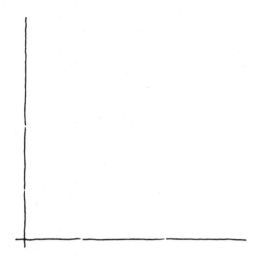

1. Sketch a right (90 deg.) angle. Use construction lines.

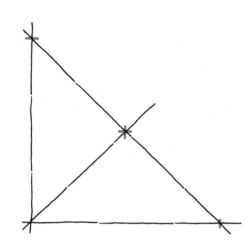

3. Divide this line into two equal parts. Starting from the point where the legs of the angle intersect, sketch a line through the dividing point of the diagonal line.

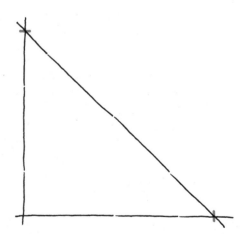

2. Units equal to the length of the desired radius are marked on each leg of the angle. Connect these points with a construction line.

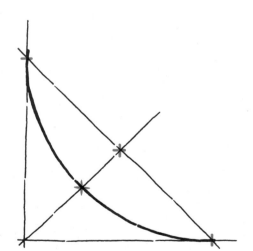

4. Mark off a point halfway between the diagonal line and the intersection of the legs of the angle. Sketch an arc through the three points as shown.

HOW TO SKETCH AN ELLIPSE

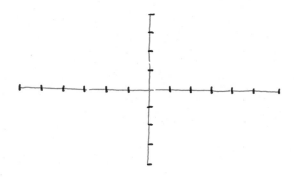

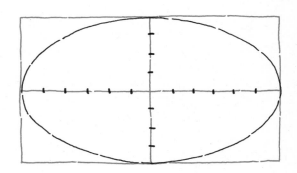

1. Sketch horizontal and vertical lines as shown. Mark off equal size units on the center lines to construct a rectangle with the dimensions equal to the major axis (the long axis) and the minor axis (the small axis) of the desired ellipse.

3. Lightly sketch arcs tangent to the lines that form the rectangle.

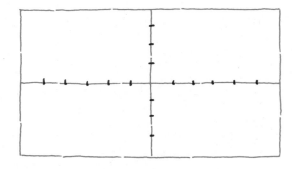

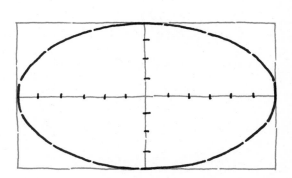

2. Construct the rectangle by sketching construction lines through the outer points.

4. When you are satisfied with the shape of the ellipse, complete it by going over the construction lines with lines of the desired weight.

HOW TO SKETCH A HEXAGON

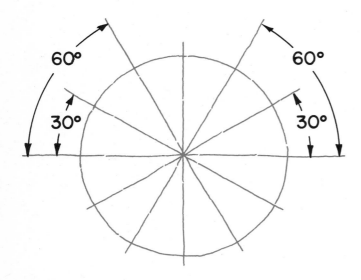

1. Sketch vertical and horizontal center lines, and inclined lines at 30 and 60 deg. Construct a circle with a diameter equal to distance across flats of required hexagon. Use construction lines.

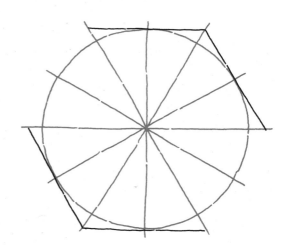

3. Sketch inclined parallel lines at 60 deg. and tangent to the circle at the point where the 30 deg. inclined line intersects the circle.

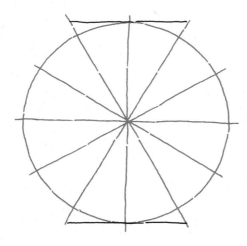

2. Sketch horizontal parallel lines at right angles (90 deg.) to the vertical center line. The lines are tangent to the circle at these points.

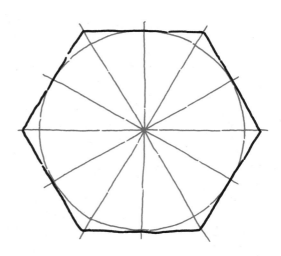

4. Complete the hexagon and go over the construction lines to produce the proper weight line.

HOW TO SKETCH AN OCTAGON

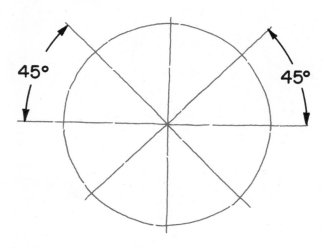

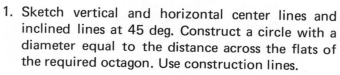

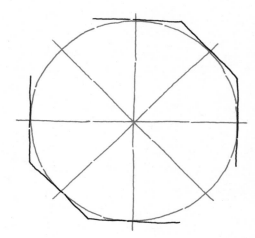

1. Sketch vertical and horizontal center lines and inclined lines at 45 deg. Construct a circle with a diameter equal to the distance across the flats of the required octagon. Use construction lines.

3. Sketch inclined parallel lines at 45 deg. and tangent to the circle at the point where the 45 deg. inclined lines intersect the circle.

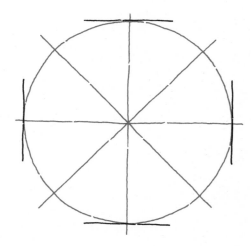

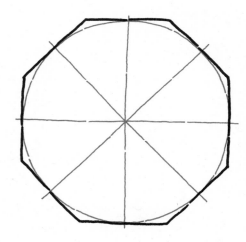

2. Sketch parallel lines tangent to the circle where the horizontal and vertical center lines intersect the circle.

4. Complete the octagon and go over the construction lines to produce the desired weight line.

SHEET LAYOUT FOR SKETCHING

1. Sketch a 15 mm border around the edges of the paper. Use a construction line. Sheet size may be Type A (8 1/2 x 11 in.) or ISO A4 (210 x 297 mm). It may be plain or graph paper. Sketch in guide lines as shown in Fig. 2-5.

2. The edge of your drawing board or desk may be used as a guide in sketching the border and guide lines, Fig. 2-6. Place the pencil in a fixed position and move your fingers along the edge of the drawing board or desk.

3. Sketch a border line over the construction lines, letter in information as shown, Fig. 2-7, or as specified by your instructor.

4. Take your time and sketch in the border and information carefully and neatly.

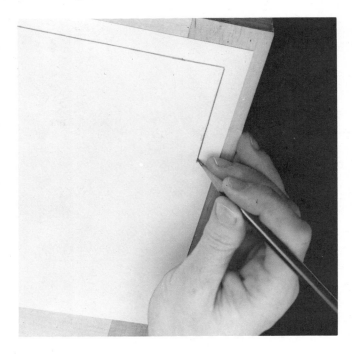

Fig. 2-6. The edge of your drawing board can be used as a guide when sketching border lines. Note how the pencil is held.

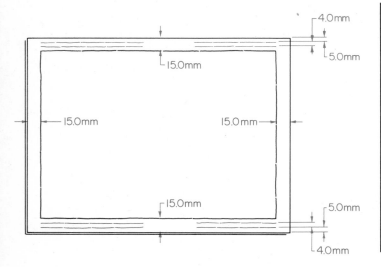

Fig. 2-5. Sketching in border and lettering guide lines.

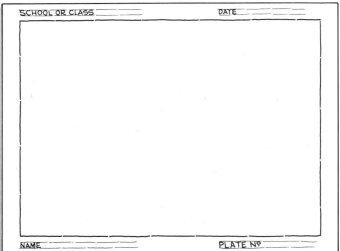

Fig. 2-7. Lettering in the required information.

ENLARGING OR REDUCING BY THE GRAPH METHOD

Drawings can be reduced or enlarged easily and rapidly by using this technique. The original drawing is blocked off into squares, Fig. 2-8. After deciding how much larger or smaller the new drawing is to be made, draw squares of the new size on a blank sheet of paper. Using the design in the small squares as a guide, sketch the design into the larger squares.

EXAMPLE: A drawing of the design is to be enlarged to twice its original size. Mark off the original drawing in 5.0 mm squares (square size will vary depending on the size of the design). Number the squares starting at the upper left side and across the top. Make up another sheet with squares that are twice the size of the 5.0 mm squares, or 10.0 mm squares. Number the large squares in the same manner

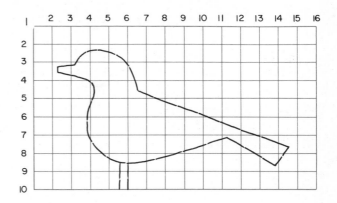

Fig. 2-8. Drawing to be enlarged has been blocked in.

as the small squares. Mark on the larger squares the points where the drawing crosses the squares. Then, sketch in the details freehand, Fig. 2-9.

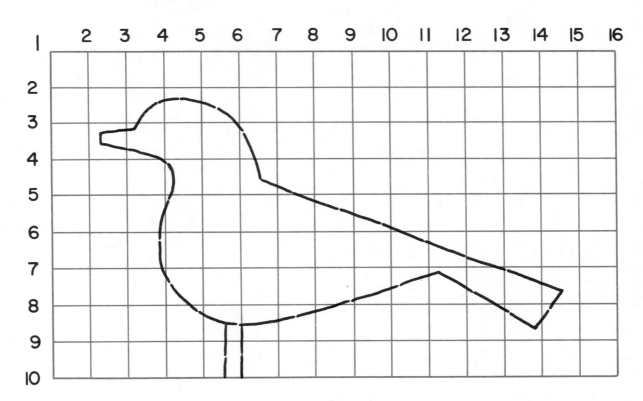

Fig. 2-9. Enlarged drawing.

TEST YOUR KNOWLEDGE - UNIT 2

(Write answers on a separate sheet of paper.)

1. In sketching, a line is drawn by making a series of _____.

2. The heaviest line used in sketching is the _____ line.

3. Drawings can be _____ or _____ easily by using the graph method.

4. When sketching an inclined line, sketch up when line inclines to _____ and down when line inclines to _____.

5. Extension lines are the same weight as _____ lines.

6. In sketching, horizontal lines are drawn from _____; vertical lines from the _____.

7. Dimension lines generally terminate in _____ at the ends.

Shop sketching is one of many drafting techniques. Drafters are required to produce drawings ranging from a simple freehand sketch to working drawings for a complex product. (AiResearch Mfg. Co.)

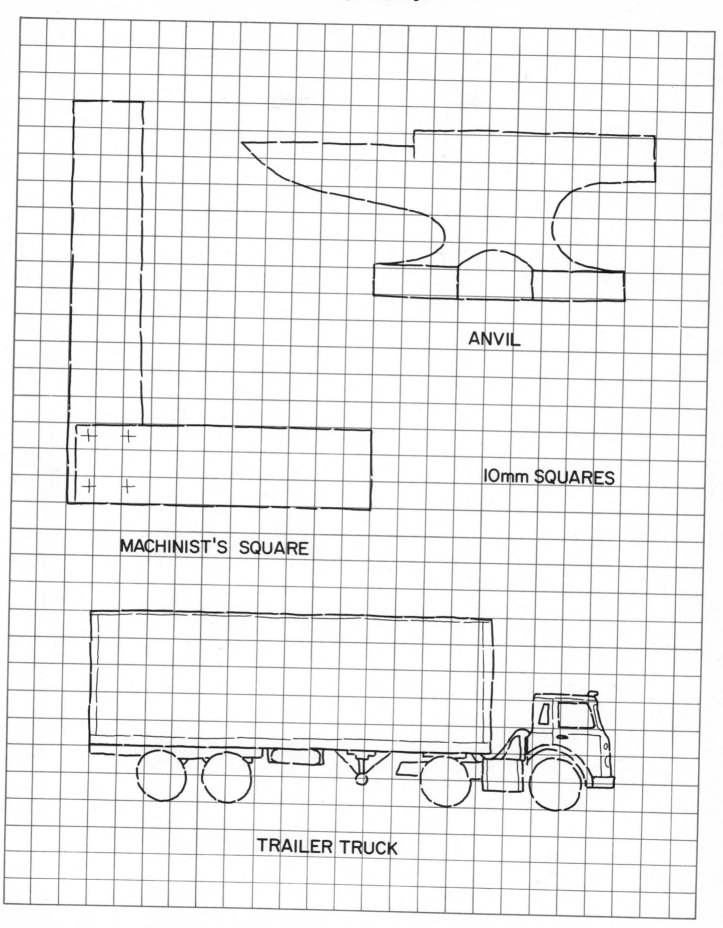

ANVIL

10mm SQUARES

MACHINIST'S SQUARE

TRAILER TRUCK

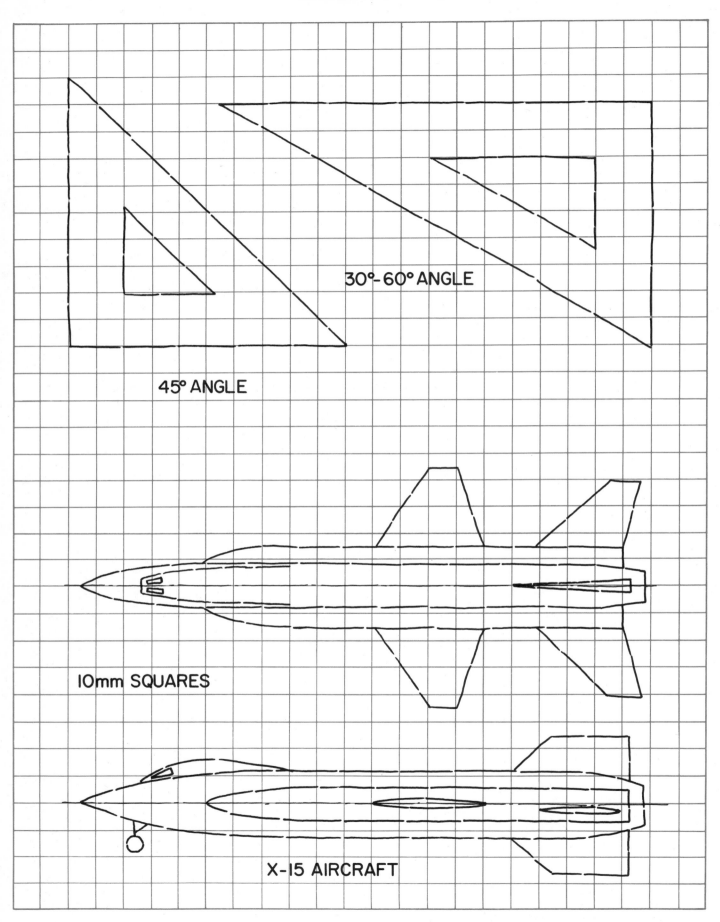

30°-60° ANGLE

45° ANGLE

10mm SQUARES

X-15 AIRCRAFT

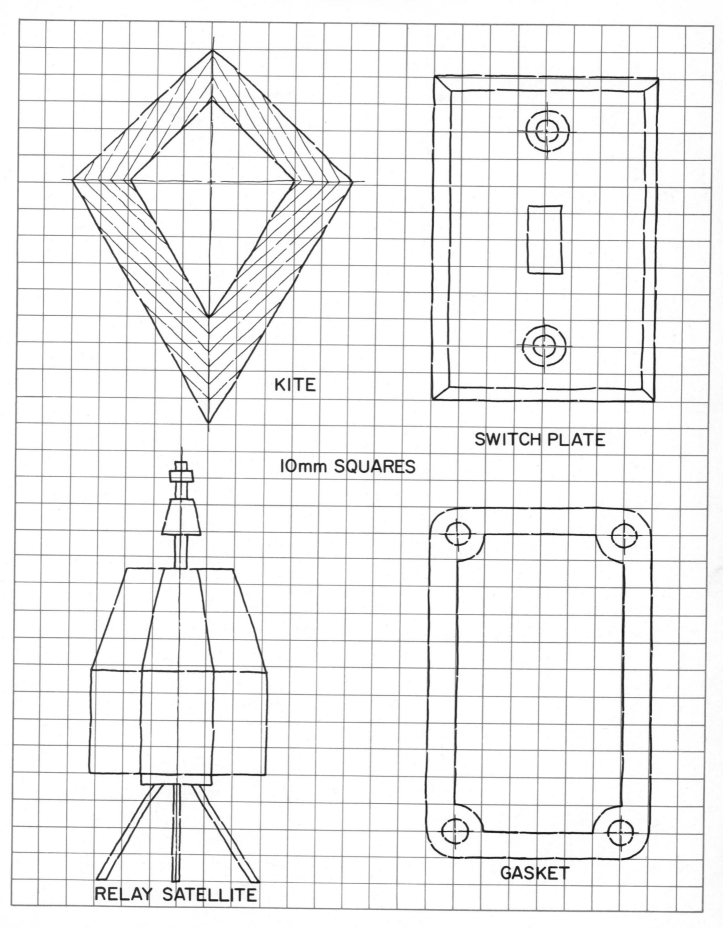

KITE

SWITCH PLATE

10mm SQUARES

RELAY SATELLITE

GASKET

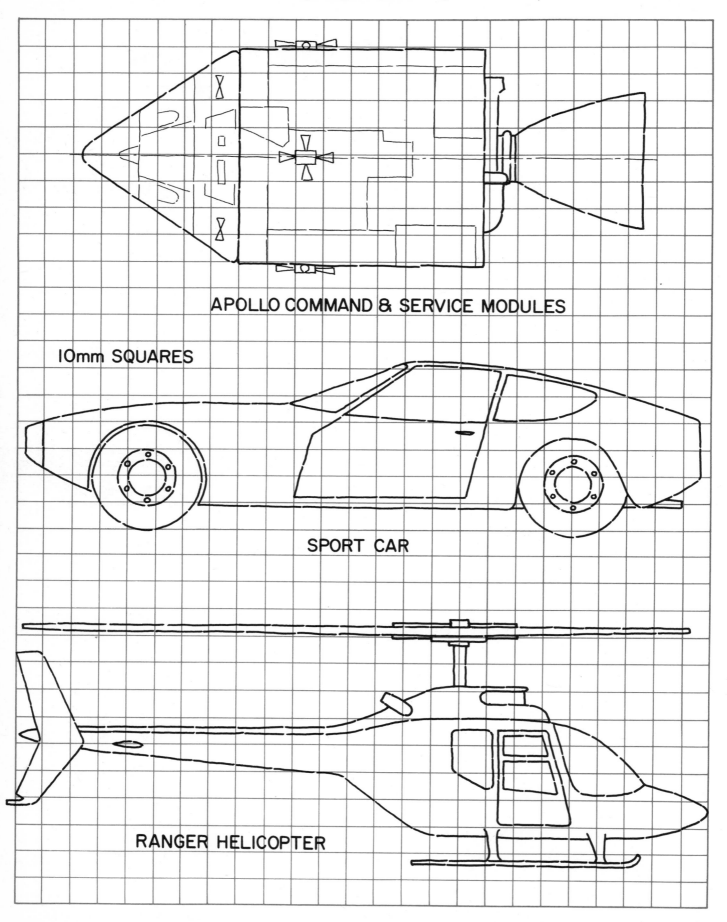

APOLLO COMMAND & SERVICE MODULES

10mm SQUARES

SPORT CAR

RANGER HELICOPTER

Unit 3
DRAFTING TOOLS

DRAWING BOARD

The DRAWING BOARD, Fig. 3-1, provides the smooth, flat surface needed for drafting. The tops of many drafting tables are designed for this purpose. Individual drawing boards are manufactured in a variety of sizes. The majority of them are made from selected, seasoned basswood.

The right-handed drafter will use the left edge of the board as the working edge; the left-handed drafter the right edge. The working edge should be checked periodically for straightness.

Drafters often tape a piece of heavy paper or a special vinyl board cover to the working face of the drawing board to protect its surface. The vinyl surface is easily cleaned.

T-SQUARE

Horizontal lines are drawn with the T-SQUARE, Fig. 3-2. It also supports triangles when they are used to draw vertical and inclined lines.

The T-square consists of two parts, the head and the blade or straight edge. The head is usually fixed solidly to the blade. However, a T-square with a protractor head and adjustable blade is also available.

Clear plastic strips inserted in the blade edge of some T-squares makes it easier to locate reference

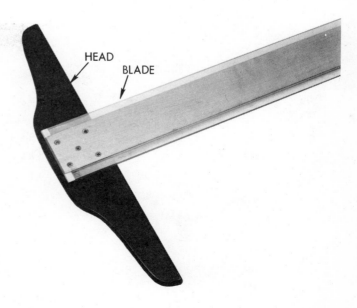

HEAD
BLADE

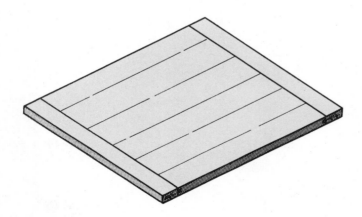

Fig. 3-1. Drawing board.

Fig. 3-2. T-square.

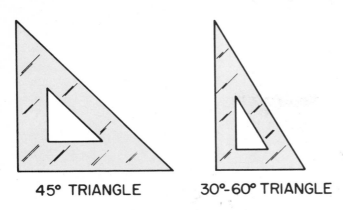

45° TRIANGLE   30°-60° TRIANGLE

Fig. 3-3. Triangles.

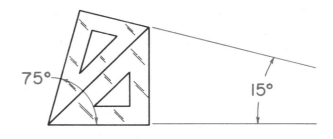

75°   15°

Fig. 3-4. Drawing 75 deg. and 15 deg. angles.

points and lines. The blade must never be used as a guide for a knife or other cutting tool.

If accurate line work is to be done, it is essential that the head of the T-square be held firmly against the working edge of the board.

It is recommended that the blade be left flat on the board or suspended from the hole in its end. This will keep warping or bowing of the blade to a minimum.

TRIANGLES

When supported on the T-square blade, the 30-60 deg. and 45 deg. TRIANGLES, Fig. 3-3, are used to draw vertical and inclined lines. They are made of transparent plastic in a number of different sizes.

To prevent warping, the triangle should be left flat on the drawing board when not being used.

Fig. 3-5. Drafting machine.   (Post)

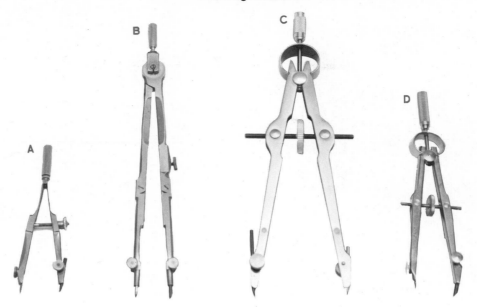

Fig. 3-6. Compasses. A—Spring bow. B—Friction type compass. C—Big bow compass. D—Small bow compass.

Angles of 15 and 75 deg. can be drawn by combining the triangles as shown in Fig. 3-4.

To draw vertical lines accurately, rest the triangle solidly on the T-square blade while holding the T-square head firmly against the working edge of the drawing board.

DRAFTING MACHINES

Industry makes considerable use of DRAFTING MACHINES, Fig. 3-5. This device replaces both the T-square and triangles. The straightedges can be adjusted to any angle. Drafting machines are gradually replacing T-squares and triangles in school drafting rooms.

COMPASS

In drafting, circles and arcs are drawn with a COMPASS, Fig. 3-6. The tool is held as shown in Fig. 3-7. For best results the lead should be adjusted so that it is about 1.0 mm shorter that the needle. Both legs will be the same length when the needle penetrates the paper. Fit the compass with lead that is one grade softer than the pencil used to make the drawing. The lead must be kept sharp.

Several attachments are available for use with the compass. A pen is substituted when inking is to be done, Fig. 3-8. An extension makes it possible to draw larger circles. However, a BEAM COMPASS will do the job better, Fig. 3-9.

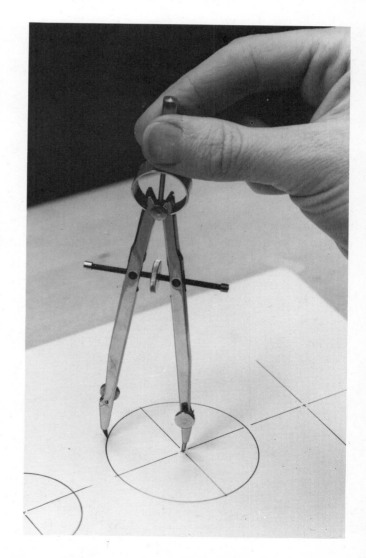

Fig. 3-7. The correct way to hold the compass when drawing a circle.

35

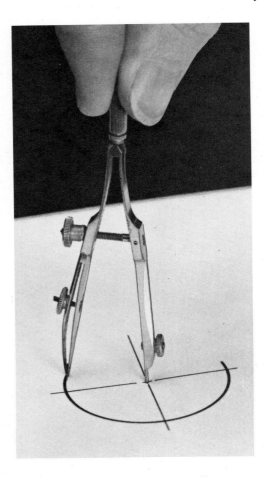

Fig. 3-8. Inking a circle.

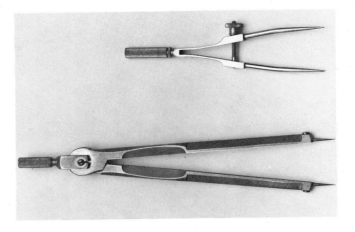

Fig. 3-10. Dividers.

DIVIDERS

Distances are subdivided and measurements are transferred with DIVIDERS, Fig. 3-10. Careful adjustment of the divider points is necessary. Also, be careful where you place the dividers or compass after use. It is very painful to accidentally run the point into your hand.

PENCIL POINTER

It is not necessary to resharpen your drawing pencil every time it starts to dull. It can be repointed quickly with a PENCIL POINTER, Fig. 3-11. Use the pencil sharpener only when the point becomes very blunt, or when it breaks.

Many commercial pencil pointers are available. The sandpaper pad, Fig. 3-12, is most frequently

To set the compass to size, draw a line on a piece of clean scrap paper and measure off the required radius. Set the compass on this line. Avoid setting a compass on a scale. "Sticking" the compass needle into the scale will eventually destroy its accuracy.

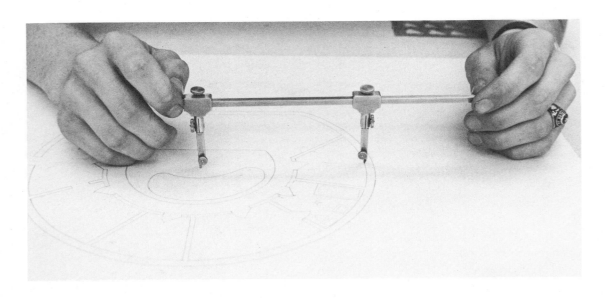

Fig. 3-9. Using a beam compass to draw large circles.

Fig. 3-11. Mechanical pencil pointers.

Fig. 3-13. A few of the many different kinds of erasers used in the drafting room.

found in the school drafting room. A piece of Styrofoam cemented to the back of the pad is used to wipe graphite dust from the freshly pointed pencil.

Keep the pencil pointer clear of the drawing area when repointing your pencil. The graphite dust will smudge your paper when you attempt to wipe it off.

ERASERS

Many shapes and kinds of ERASERS, Fig. 3-13, are manufactured for use in the drafting room. The type of material being drawn upon — paper, film or vellum, will determine the kind of eraser to be used.

Brush away eraser crumbs before starting to draw again.

ERASING SHIELD

Small errors or drawing changes can be erased without soiling a large section of the drawing if an ERASING SHIELD is employed, Fig. 3-14. Place an opening in the shield of the proper shape and size over the area to be changed and then erase. The erasure is made without touching other parts of the drawing.

Fig. 3-12. Using a sandpaper pad to point a drafting pencil.

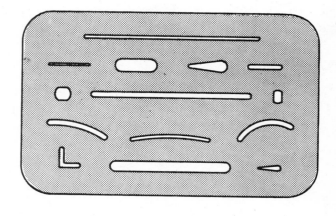

Fig. 3-14. Erasing shield.

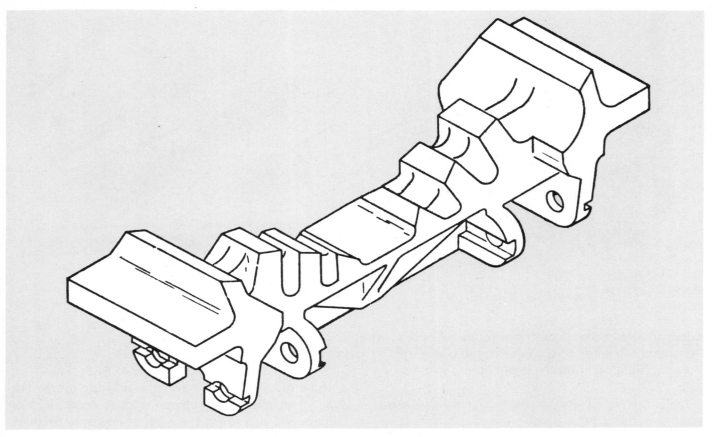

Above is an illustrator's isometric interpretation of a complex casting used in logging equipment, which took 7.6 hours to prepare. Below is a drawing of the same casting which an automated drafting machine made in 3/4 of an hour. (Perspective Inc.)

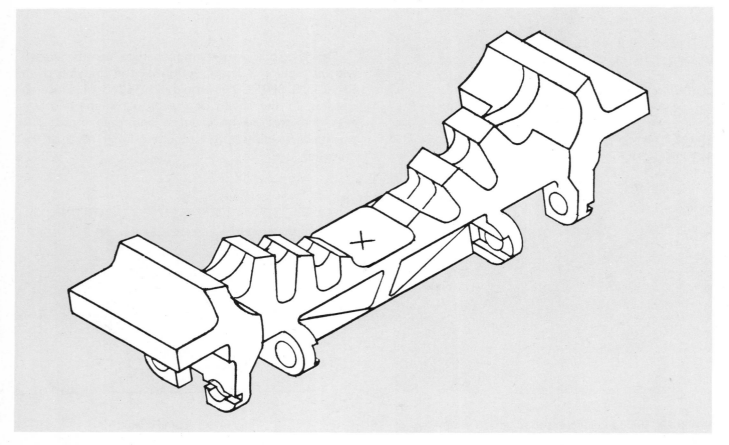

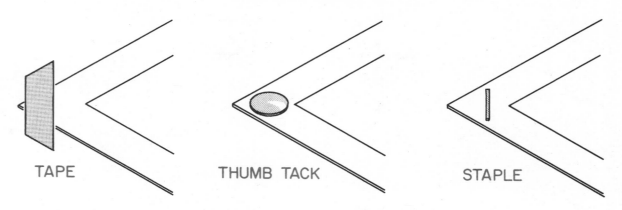

TAPE

THUMB TACK

STAPLE

Fig. 3-15. Methods of attaching paper to the drawing board.

A popular shield is made from stainless steel. This metal can be made very thin and still be strong. It is also wear resistant and does not stain or "smudge" the drawing.

FASTENERS

Three methods of attaching paper to the drawing board are shown in Fig. 3-15. Tape is the most desirable of the methods. It does not puncture the paper or the drawing board.

Staples may be used but their continued use will tear up the working surface of the drawing board. They are often difficult to remove.

Thumb tacks are least recommended. They quickly destroy the smooth working surface of the board.

DUSTING BRUSH

No matter how careful you are, some erasing crumbs and dirt particles will collect on the drawing area. These should be removed by using a DUSTING BRUSH, Fig. 3-16, rather than your hands. Using your hands may cause smudges and streaks.

Dusting brushes are available in a number of sizes and with natural and manufactured bristles.

PROTRACTORS

A PROTRACTOR, Fig. 3-17, is used to measure and lay out angles on drawings. Usually, they are made of clear plastic and may be either circular or semicircular in shape. The degree graduations are scribed or engraved around the circumference of the protractor.

When measuring or laying out an angle, place the center lines of the protractor at the point of the required angle as shown in Fig. 3-18. Read or mark the angle from the graduations on the circumference of the tool.

FRENCH CURVES

Curved lines that are not exactly circular in form are drawn with a FRENCH CURVE, Fig. 3-19. After

Fig. 3-16. Removing erasure crumbs from the drawing with a dusting brush.

Fig. 3-17. Protractor.

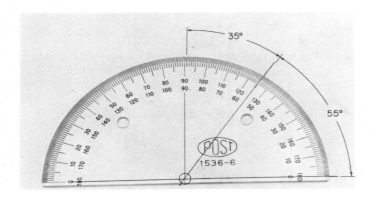

Fig. 3-18. Making a measurement with a protractor.

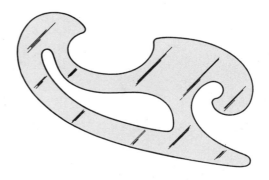

Fig. 3-19. French curve.

the curved line is carefully plotted, it is drawn using a French curve as shown in Fig. 12-6, page 150.

The curves are made of transparent acrylic plastic. They range in size from a few inches to several feet in length and may be purchased individually or as a set.

TEMPLATES

Much time can be saved in drawing standard symbols and figures if TEMPLATES (patterns or guides) are used, as shown in Fig. 3-20. Made of thin plastic, these tools are available in a large number of styles and sizes.

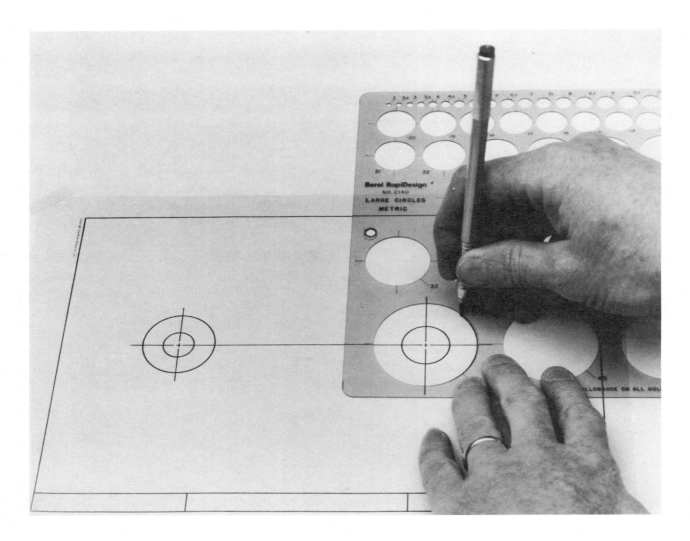

Fig. 3-20. Much time can be saved by using a template.

DRAFTING MEDIA

Drawings are made on many different materials — paper, tracing vellum, film, etc. A heavyweight opaque paper that is white, buff or pale green in color is used in many school drafting rooms.

While this paper takes pencil lines well, it is difficult to erase because the pencil point makes a depression in the paper when a line is drawn. If this material is used, take special care to prevent mistakes.

Industry makes much use of tracing vellum and film because reproductions or prints must be made of all drawings.

Drawing media is available in either sheet or roll form. Any size in the A series can be obtained by enlarging and reducing, and it will retain the same proportion.

SIZE	MILLIMETRES	INCHES
A4	210 x 297	8.27 x 11.69
A3	297 x 420	11.69 x 16.54
A2	420 x 594	16.54 x 23.39
A1	594 x 841	23.39 x 33.11
A0	841 x 1189	33.11 x 46.81

PENCILS

As most drawings are prepared with a pencil, it is important that the proper pencil be selected. The drawing media used will determine the type of pencil that should be used. The conventional graphite lead pencil is satisfactory with most papers and tracing vellums, while a pencil with plastic lead is necessary if the drawing is made on film.

The drafter can select from 17 grades of pencils that range in hardness from 9H (very hard) to 6B (very soft), Fig. 3-21. Many drafters use a 4H or 5H pencil for layout work and a H or 2H pencil to

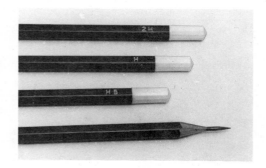

Fig. 3-21. Pencils used in drafting are available in a wide range of hardness.

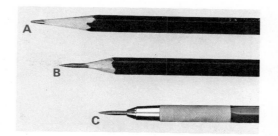

Fig. 3-22. Pencil points. A—Sharpened with regular pencil sharpener. B—Sharpened with a knife and sandpaper pad. C—Sharpened with mechanical pencil pointer.

darken the lines and to letter. In general, use a pencil that will produce a sharp, dense black line because this type of line reproduces best on prints.

Avoid using a pencil that is too soft. It will wear rapidly, smear easily and soil your drawing. Also, the lines will be "fuzzy" and will not produce usable prints.

A conical shaped pencil point is preferred for most general purpose drafting. To sharpen the pencil, cut the wood away from the unlettered end. Use a knife or mechanical sharpener and point the lead on a pencil pointer, Fig. 3-22.

A semiautomatic pencil, Fig. 3-23, is usually preferred to a wood pencil. With this type pencil, it is not necessary to cut away the wood to expose the lead. The extended lead is shaped on a pencil pointer.

SCALES

SCALES, Figs. 3-24 and 3-25, are in constant use on the drawing board because almost every line on a mechanical drawing must be of a measured length. Accurate drawings require accurate measurements.

Because of the diversity of work to be drawn, scales used by the drafter are made in many shapes, lengths and measurement graduations. Scales may be made of wood, plastic or a combination of both materials. Graduations are printed on inexpensive scales — machine-engraved on the more costly ones.

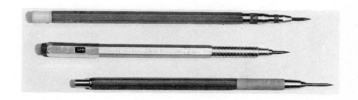

Fig. 3-23. A variety of semiautomatic drafting pencils.

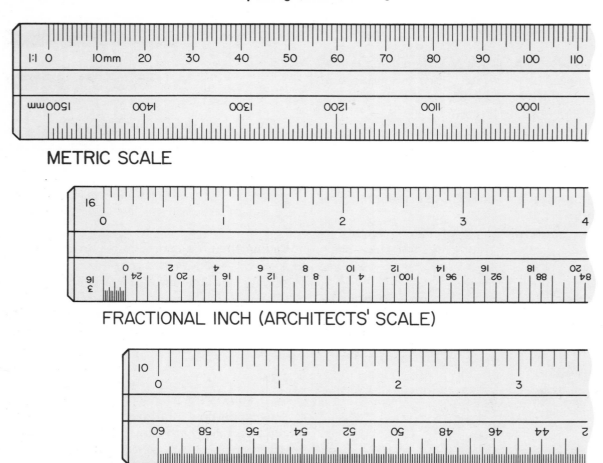

METRIC SCALE

FRACTIONAL INCH (ARCHITECTS' SCALE)

DECIMAL INCH (ENGINEERS' SCALE)

Fig. 3-24. Drafting scales.

See Fig. 3-24 for a comparison of a simple metric scale to the more conventional fractional inch and decimal inch scales.

The METRIC SCALE, now a required tool of the trade, is marked in millimetres (mm). See Fig. 3-25. Numbered lines on a metric scale used for full size drawings (1:1 ratio) are marked 10, 20, 30, etc. Avoid scales marked in centimetres (cm). In addition to the full size scale (1:1 ratio), metric scales are available in a number of enlargement ratios (2:1, 3 1/3:1, etc.) and reduction ratios (1:2, 1:3, etc.).

Use a suitable metric scale when making metric dimensioned drawings. AVOID converting metric dimensions to inches and fractions, and using a conventional scale to make measurements. THINK METRICS. FORGET ABOUT INCHES AND FRACTIONS.

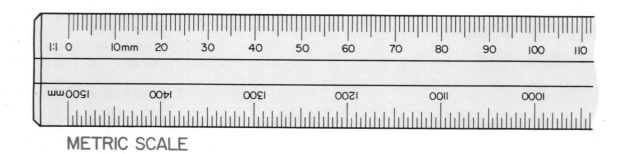

METRIC SCALE

Fig. 3-25. Metric scale. Upper edge is for full size drawings and has a 1:1 ratio. Bottom edge has a 1:5 reduction ratio (one-fifth full size).

TEST YOUR KNOWLEDGE - UNIT 3

(Write answers on a separate sheet of paper.)

1. The drawing board provides the _____
   _____.

2. A piece of heavy drawing paper or a special vinyl board cover is sometimes attached to the working surface of the board to:
   a. Provide a drawing surface.
   b. Protect its surface.

3. The T-square is used to _____.

4. Vertical and inclined lines are drawn with _____.

5. The compass is used in drafting to draw _____.

6. _____ is used to repoint a pencil.

7. The erasing shield is used to _____
   _____.

8. The three methods used to attach drawing paper to the drawing board are _____.
   Which method least damages the board?

9. Angles can be measured and laid out on drawings by using a _____.

10. The _____ or _____ pencil is recommended for layout work.

11. An _____ or _____ pencil is used to darken the lines and for lettering.

12. Scales are important to drafters because _____
    _____.

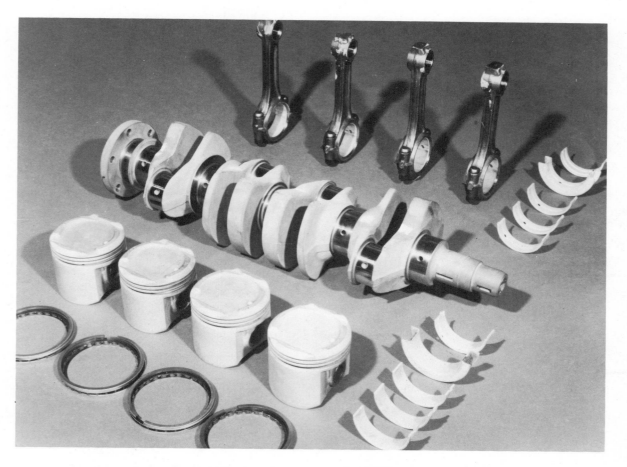

Automotive engine components built to metric standards and assembled with metric fasteners are becoming more commonplace.   (Chevrolet Motor Div., General Motors Corp.)

Unit 4
DRAFTING TECHNIQUES

You were introduced to the ALPHABET OF LINES in the Unit on Sketching. The lines you sketched can be drawn more uniformly and accurately with drafting instruments. See Fig. 4-1.

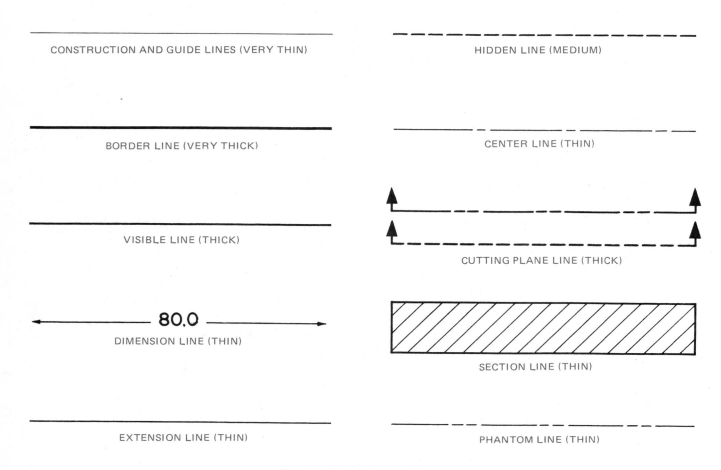

CONSTRUCTION AND GUIDE LINES (VERY THIN)

HIDDEN LINE (MEDIUM)

BORDER LINE (VERY THICK)

CENTER LINE (THIN)

VISIBLE LINE (THICK)

CUTTING PLANE LINE (THICK)

80.0
DIMENSION LINE (THIN)

SECTION LINE (THIN)

EXTENSION LINE (THIN)

PHANTOM LINE (THIN)

Fig. 4-1. The alphabet of lines.

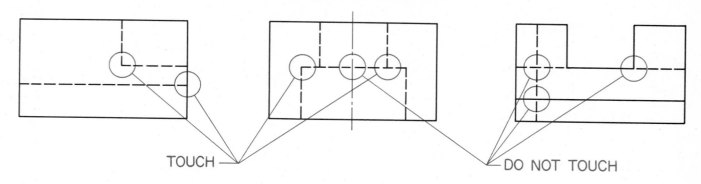

Fig. 4-2. Correct use of hidden lines.

To understand the LANGUAGE OF INDUSTRY, it is necessary that you know the characteristics of the various lines and the correct way to use them in a drawing, Fig. 4-2. In drafting room language the characteristics and uses of lines are known as LINE CONVENTIONS.

Each type or kind of line has a specific meaning. It is most important that each line is drawn properly, is opaque and uniform in width throughout its entire length. See Fig. 4-3.

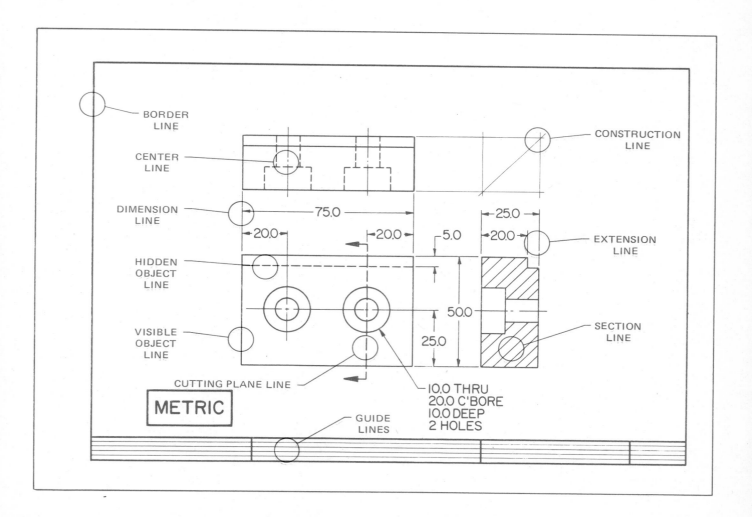

Fig. 4-3. How the various lines are used.

ALPHABET OF LINES

CONSTRUCTION AND GUIDE LINES (VERY THIN)

CONSTRUCTION LINES are drawn very lightly. They are used to block in drawings and as guide lines for lettering. They may be erased, if necessary, after they have served their purpose.

BORDER LINE (VERY THICK)

The BORDER LINE is the heaviest weight line used in drafting. It varies from 1.0 to 2.0 mm thick, depending upon the size of the drawing sheet.

VISIBLE LINE (THICK)

The VISIBLE OBJECT LINE (also VISIBLE LINE) is used to outline the visible edges of the object being drawn. They should be drawn so that the views stand out clearly on the drawing. All of the visible object lines on the drawing should be the same weight.

80.0

DIMENSION LINE (THIN)

The DIMENSION LINE usually is capped at each end with arrowheads and is placed between two extension lines. With few exceptions, it is broken with the dimension placed at midpoint between the arrowheads. The dimension line is a light line, a bit heavier than the construction line. It is placed a minimum of 10.0 mm away from the drawing.

EXTENSION LINE (THIN)

The EXTENSION LINE is the same weight as the dimension line. It extends the dimension line beyond the outline of the view so that the dimension can be read easily. The line starts about 1.5 mm away from the view and extends about 1.5 mm past the last dimension line.

HIDDEN LINE (MEDIUM)

The HIDDEN OBJECT LINE (also HIDDEN LINE) is used to show the hidden features of the object. It is drawn a bit lighter than the visible object line. It is made up of broken lines with dashes about 4.0 mm long with about 1.0 mm spacing. They may vary slightly according to size of drawing. Hidden object lines should always start and end with a dash in contact with the visible object line. See Fig. 4-2 for the correct uses of the hidden object line.

CENTER LINE (THIN)

The CENTER LINE is used to indicate the center of symmetrical objects. It is a fine, dark line composed of alternate long dashes (10 to 40 mm long) and short dashes and spaces (about 3.0 mm and 1.5 mm respectively) between. The center line should extend uniformly only a short distance beyond the circle or view. They start and end with long dashes and should not cross at the spaces between the dashes.

CUTTING PLANE LINE (THICK)

The CUTTING PLANE LINE is slightly heavier than the visible line. It is used to indicate where the sectional view has been taken. Two types are acceptable. However, a cutting plane line made of 6.5 mm dashes with 1.0 mm spacing is recommended. The second type uses a long dash (15-40 mm and short 3.0 mm dashes). The cutting plane line will be further explained in Unit 9 on Sectional Views.

SECTION LINE (THIN)

SECTIONAL LINES are used when drawing the inside features of the object. They indicate material cut by the cutting plane line, and also indicate the general classification of the material. The lines are fine dark lines.

PHANTOM LINE (THIN)

The PHANTOM LINE is used to indicate alternate positions of moving parts and outlines of related parts. It is a thin dark line made of long dashes (15 to 40 mm long), alternated with pairs of short dashes 2.0 to 3.0 mm in length, with 1.0 mm spaces.

HOW TO MEASURE

Almost every line on a mechanical drawing must be a measured line. If your drawings are to be made accurately, you must be able to make accurate measurements.

Measurements are made in the drafting room with SCALES, Fig. 4-4. The term SCALE is used to indicate both the device used to measure and the size to which an object is to be drawn.

Scales have graduations on the edges which show lengths used to indicate smaller or larger units of measure, Fig. 4-5.

Several different shapes of metric scales are available, Fig. 4-6.

A SCALE CLIP, Fig. 4-7, provides a means to lift a triangular scale. It keeps the desired scale edge in an upright position.

There are no architects' or engineers' scales with the metric system. Most drawings will be dimensioned

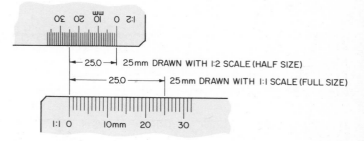

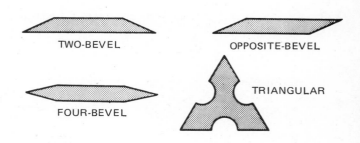

Fig. 4-5. The graduations on the scale edges indicate larger or smaller scale units of measure.

TWO-BEVEL OPPOSITE-BEVEL

FOUR-BEVEL TRIANGULAR

Fig. 4-6. Several different shapes of metric scales are available.

Fig. 4-7. The scale clip keeps the desired scale edge in an upright position.

in millimetres. Full size drawings will require a 1:1 ratio scale. Use a 1:2 ratio scale when drawings must be reduced to half size. The preferred metric scales are shown in Fig. 4-8. For architectural drafting, reduction scales above 1:10 are primarily used. See Unit 21 for more details.

COMMON DRAFTING SCALES

CUSTOMARY INCH		NEAREST ISO METRIC EQUIVALENT (mm)
1:2500	(1 in. = 200 ft.)	1:2000
1:1250	(1 in. = 100 ft.)	1:1000
1:500	(1/32 in. = 1 ft.)	1:500
1:192	(1/16 in. = 1 ft.)	1:200
1:96	(1/8 in. = 1 ft.)	1:100
1:48	(1/4 in. = 1 ft.)	1:50
1:24	(1/2 in. = 1 ft.)	1:20
1:12	(1 in. = 1 ft.)	1:10
1:4	Quarter size (3 in. = 1 ft.)	1:5
1:2	Half-size (6 in. = 1 ft.)	1:2
1:1	Full-size (12 in. = 1 ft.)	1:1

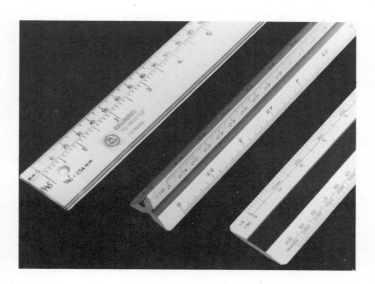

Fig. 4-4. Various types of metric scales for drafting.

Fig. 4-8. Preferred metric scales recommended by ISO.

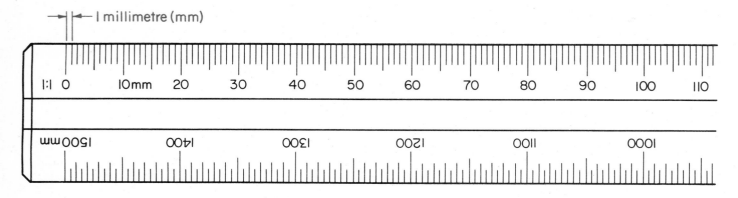

Fig. 4-9. Section of 1:1 metric scale. Each division is equal to 1.0 mm.

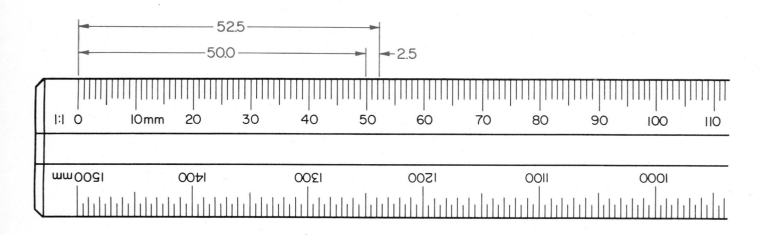

Fig. 4-10. Making a measurement of 52.5 mm.

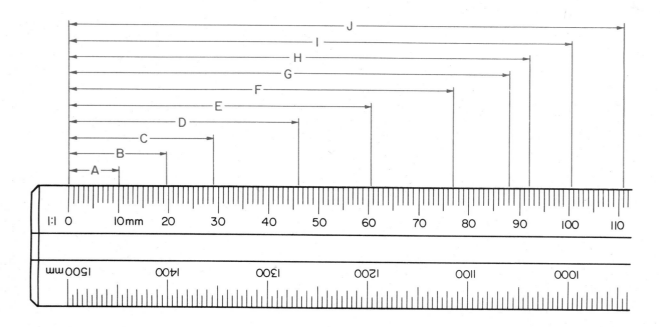

Fig. 4-11. How many can you answer correctly? On a piece of paper, write the letters A to J. After each letter, write the correct measurement. Ask your instructor to check your answers.

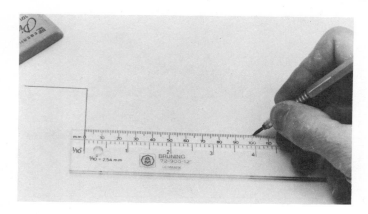

Fig. 4-12. When making a measurement, observe the scale directly from above.

Most of the drawings you will make are measured with a 1:1 scale. Each division is 1.0 mm, and the numbered lines are 10, 20, 30, etc., Fig. 4-9.

To make a measurement of 52.5 mm, for example, it is a simple matter to come out to the 50 division, then add 2.5 more, Fig. 4-10.

A section of a 1:1 scale is shown in Fig. 4-11. How many of the measurements can you read correctly?

MAKING MEASUREMENTS

To make a measurement, observe the scale from directly above. Mark the desired measurement on the paper by using a light perpendicular line made with a sharp pencil, Fig. 4-12.

Fig. 4-13. Use the T-square to draw horizontal lines. Note how the T-square head is held firmly in place against the edge of the drawing board.

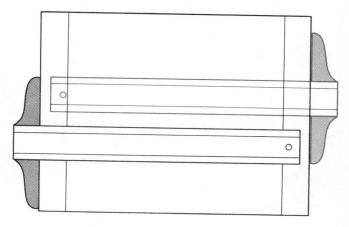

Fig. 4-14. Left-handed drafters use the RIGHT edge of the drawing board.

Keep the scale clean. Do not mark on it, or use it as a straightedge.

HOW TO DRAW LINES
WITH INSTRUMENTS

Care must be taken if lines are to be the same weight; that is, uniform in width and darkness.

When lines are drawn using instruments, hold the pencil perpendicular to the paper, and inclined at an angle of about 60 deg. in the direction the line is being drawn. To keep the lines uniform in weight, especially if a long line is being drawn, rotate the pencil as you draw. Rotating the pencil will also keep the point sharp.

The T-square is used to make horizontal lines, Fig. 4-13. The lines are drawn from left to right. Hold the T-square head firmly against the LEFT edge of the drawing board (left-handed drafters will use the RIGHT edge of the board, Fig. 4-14).

Vertical lines may be drawn using a triangle and are drawn from the bottom to the top of the sheet, Fig. 4-15. The base of the triangle must rest on the blade of the T-square. Use the left hand to hold the triangle in place.

The procedure to follow in drawing inclined lines (lines that are not vertical or horizontal) depends on the direction of the slope. Lines that incline to the left, Fig. 4-16, are drawn more easily from the top of the sheet down. Those that incline to the right, Fig. 4-17, should be drawn from the bottom of the sheet up. This follows the natural tendency for a person to read or write from left to right.

Fig. 4-15. Vertical lines are drawn using the triangle. The lines are drawn from the bottom to the top of the sheet.

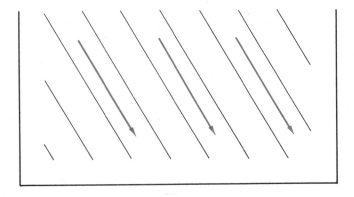

Fig. 4-16. Drawing lines that slant to the left.

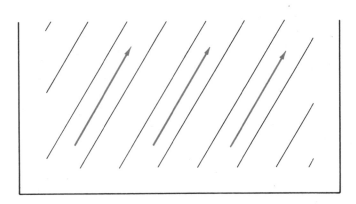

Fig. 4-17. Lines that incline to right . . . from bottom up.

HOW TO DRAW A LINE PERPENDICULAR TO A GIVEN LINE USING INSTRUMENTS

To draw a line perpendicular to a given line using instruments, place the hypotenuse (long edge) of any triangle parallel to the given line, Fig. 4-18. Support the angle on the T-square or another triangle. Rotate the first triangle around the 90 deg. corner to draw a line perpendicular to the given line.

Another technique used to draw a line perpendicular to a given line requires that you place either leg of the triangle parallel to the given line, Fig. 4-19. Support it on the T-square or another triangle. Slide the original triangle on the support and draw the line that will be perpendicular to the given line.

HOW TO ERASE

When drawing, every effort must be made to prevent mistakes. However, even the best drafter must occasionally make changes or corrections on a drawing that will require erasing.

Some helpful suggestions to follow:
1. Keep your hands and instruments clean. This will

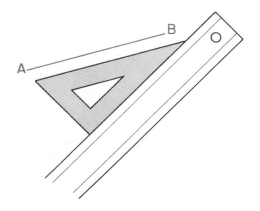

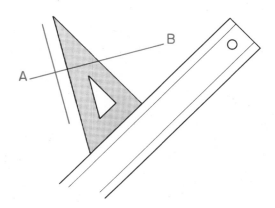

Fig. 4-18. How to draw a line perpendicular to a given line using the T-square and triangle.

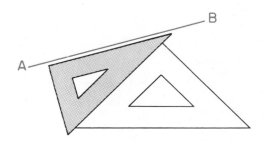

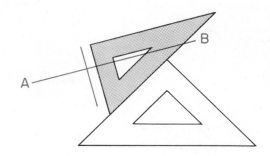

Fig. 4-19. How to draw a line perpendicular to a given line using two triangles.

help to keep "smudges" from forming.

2. Use an ERASING SHIELD, Fig. 4-20, whenever possible. Select an opening that will expose the area to be erased. The shield will protect the rest of the drawing while the erasure is made.

3. Clean the eraser crumbs from the board immediately after making an erasure. Remove them with a brush or clean cloth, not your hands.

4. Hold the paper with your free hand when erasing. This will keep the paper from wrinkling.

5. Erasing will remove the lead but it will not remove the pencil grooves from the paper. Avoid deep, wide grooves by first blocking in all views with light construction lines.

HOW TO USE A COMPASS

In general drafting work, the compass is used to draw circles and arcs. Care must be taken so that the line drawn with the compass is the same weight as the line produced with the pencil. To accomplish this, it is recommended that the compass lead should be several grades softer than that of the pencil.

Sharpen the lead and adjust the point as shown in Fig. 4-21. Do not forget to readjust the point after each sharpening.

To set the compass, draw a line that is equal in length to the desired radius. Adjust the compass on this line, Fig. 4-22. Do not set the compass on the scale because the point will eventually ruin the division lines on the scale.

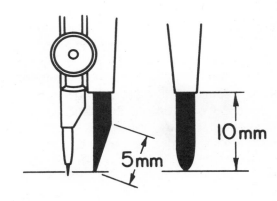

Fig. 4-21. Sharpening and adjusting the compass lead. Do not forget to readjust the point after each sharpening.

Fig. 4-20. Using the erasing shield.

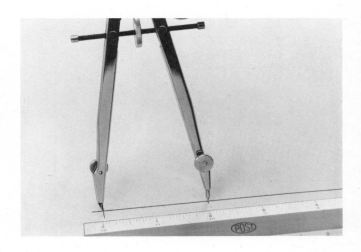

Fig. 4-22. Adjust the compass to size on a measured line. Never set it on the scale.

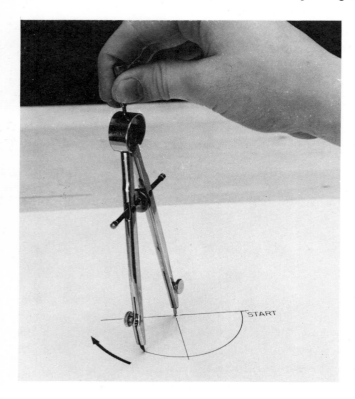

Fig. 4-23. Drawing a circle. Note that the compass is inclined in the direction of rotation.

To draw the circle, rotate the compass in a clockwise direction, Fig. 4-23, with the tool inclined in the direction of rotation. Start and complete the circle on a center line. When drawing a series of concentric circles (circles with same center), draw the smallest circle first.

Again, make sure that the circle drawn is the same line weight as the penciled line.

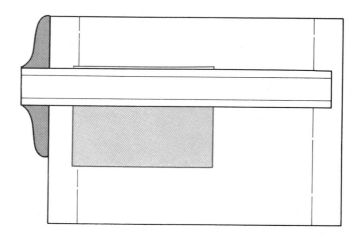

Fig. 4-25. Aligning the drawing sheet with the T-square.

ATTACHING DRAWING SHEET TO BOARD

The drawing sheet can be attached to the board with drafting tape, thumbtacks or staples. Tape is preferred because it does not damage the board.

Before attempting to attach the paper to the board, remove all eraser crumbs. To attach the paper, place the sheet on the board as shown in Fig. 4-24. Left-handed drafters should use the upper right-hand corner of the board.

Place the T-square on the board with the head firmly against the left edge. Slide it up until the top of the blade is in line with the top edge of the drawing sheet, Fig. 4-25. Position the sheet so the top edge is parallel with the T-square blade and fasten the

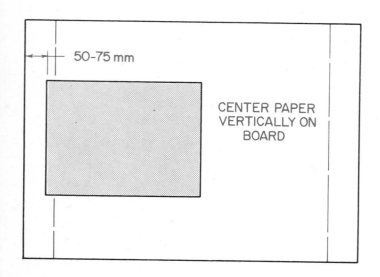

Fig. 4-24. Locating the drawing sheet on the board.

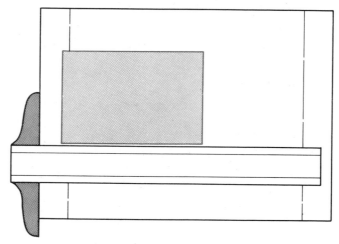

Fig. 4-26. The drawing sheet can also be aligned on the board by placing the bottom edge of the sheet on the edge of the T-square.

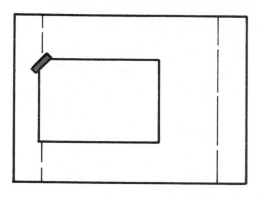

1. USE THE T-SQUARE TO LINE UP THE SHEET ON THE BOARD. FASTEN THE UPPER LEFT CORNER OF THE SHEET.

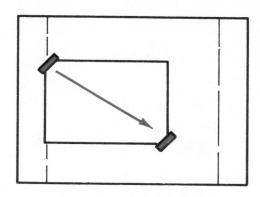

2. SMOOTH TO THE LOWER RIGHT CORNER. ATTACH SHEET.

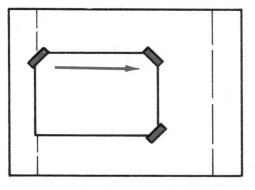

3. SMOOTH TO THE UPPER RIGHT CORNER. ATTACH SHEET.

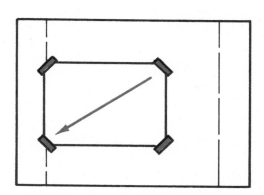

4. SMOOTH TO THE LOWER LEFT CORNER. FINISH ATTACHING THE SHEET.

Fig. 4-27. Sequence recommended for attaching lightweight papers (such as tracing vellums) to the board.

sheet to the board. Fasten bottom corners.

To position and attach A4 size (210 x 297 mm or 8 1/2 x 11 in.) drawing sheets:
1. Place the sheet on the board as shown in Fig. 4-24.
2. Place the head of the T-square firmly against the left edge of the board.
3. Align the drawing sheet by sliding the T-square blade until it contacts the bottom edge of the paper, Fig. 4-26. Align the sheet with this edge. Fasten the sheet to the board.

Lightweight paper, like tracing vellum, is slightly more difficult to attach to the board because it has a tendency to wrinkle. Align sheet on drawing board with T-square. Attach it as shown in Fig. 4-27.

DRAFTING SHEET FORMAT

Most drafting rooms use a standard format in the layout of their drawing sheets. In general, the format

consists of the border, title block and standard notes, Fig. 4-28.

The border is included to define the drawing area of the sheet. The title block and standard notes provide information that is necessary for the manufacture or assembly of the object described on the drawing sheet.

Most industrial firms use standard drawing sheet sizes. Drawings made on standard size sheets are easier to file and present less difficulty when prints are made from them.

With few exceptions, the drawings in this text can be drawn on A4 (210 x 297 mm or 8 1/2 x 11 in.) or A3 (297 x 420 mm or 11 x 17 in.) size drawing sheets.

Plan your work carefully. Do not use a large sheet if the smaller size sheet will do.

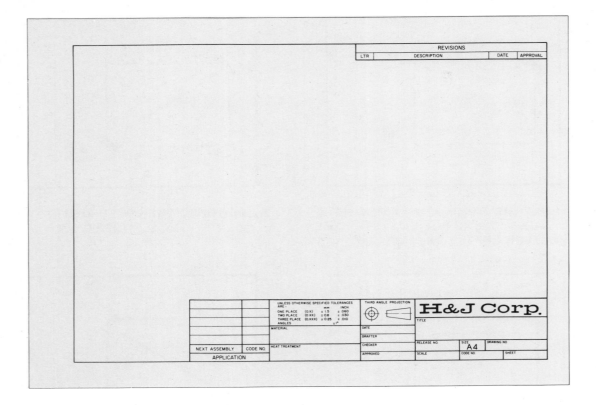

Fig. 4-28. Preprinted drawing sheets save a great deal of time
for the drafter.

RECOMMENDED DRAWING SHEET FORMAT

The following drawing sheet format is recommended for most of the problems presented in this text:

1. Place the border on the sheet as shown in Fig. 4-29. Use a short, light pencil stroke, not a dot, as a guide for drawing the border line. The short, light guide mark should be covered when the border is drawn.

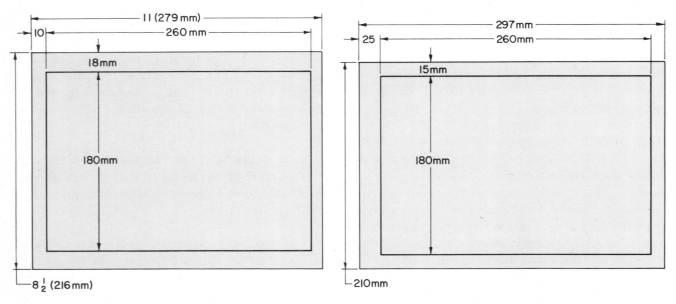

Fig. 4-29. Left. Type A sheet is a standard 8 1/2 x 11 in. sheet. Note how it is
laid out for metric drawings. Right. ISO A4 sheet shows layout for standard
metric sheets you will be using.

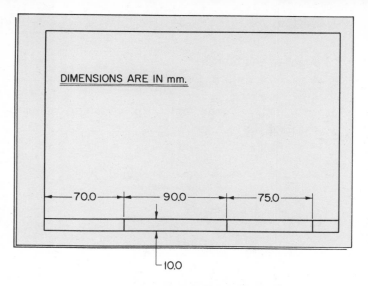

DIMENSIONS ARE IN mm.

70.0 | 90.0 | 75.0

10.0

Fig. 4-30. Horizontal drawing sheet format.

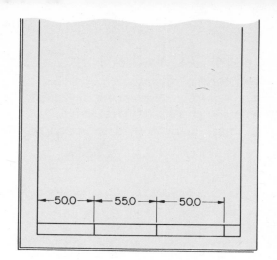

50.0 | 55.0 | 50.0

Fig. 4-31. Vertical drawing sheet format.

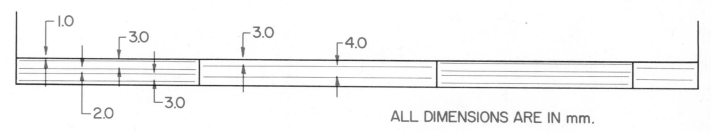

1.0
3.0
3.0
4.0
3.0
2.0

ALL DIMENSIONS ARE IN mm.

Fig. 4-32. Layout of guide lines for lettering.

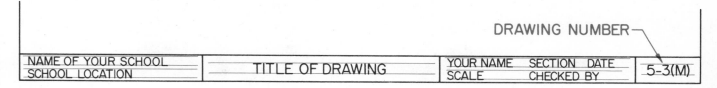

DRAWING NUMBER

NAME OF YOUR SCHOOL
SCHOOL LOCATION | TITLE OF DRAWING | YOUR NAME   SECTION   DATE
SCALE   CHECKED BY | 5-3(M)

Fig. 4-33. Information suggested to be lettered on drawing sheets for material in this text.

2. Allow 10 mm for the title block.
3. Divide the title block as shown in Figs. 4-30 and 4-31.
4. Guide lines for the title block are drawn, Fig. 4-32. The guide lines should be drawn VERY lightly.
5. Letter in the necessary information, Fig. 4-33.

TEST YOUR KNOWLEDGE - UNIT 4

(Write answers on a separate sheet of paper.)

1. Identify these lines:
   a. _____
   b. _____
   c. _____
   d. ◄——— 80.0 ———►
   e. — — — — — — — — — —
   f. ———— · — ———— · — ————

g. ↑ — — — — — — — — ↑
   ↑ — — — — — — — — ↑

h.

i. ———— — · — ———— — · —

2. In drafting room language, the characteristics of the above lines and their correct use are known as _____.
3. Why is it important for a drafter to be able to measure accurately?
4. In drafting, the term SCALE has two meanings. What are they?
5. Prepare sketches which show four different shapes of scales available.
6. In drafting, horizontal lines are drawn using the _____.

7. Vertical lines are drawn using _____.
8. When erasing, the _____ is often used to protect surrounding areas.
9. Circles and arcs are drawn with a _____.
10. List three methods used to attach the drawing

sheet to the board.

a. _____   b. _____

c. _____

11. The method used most for attaching drawing sheets is the _____.

Drafting techniques involve the use of special drafting tools and equipment.   (Bel Air Senior High School, Bel Air, Maryland)

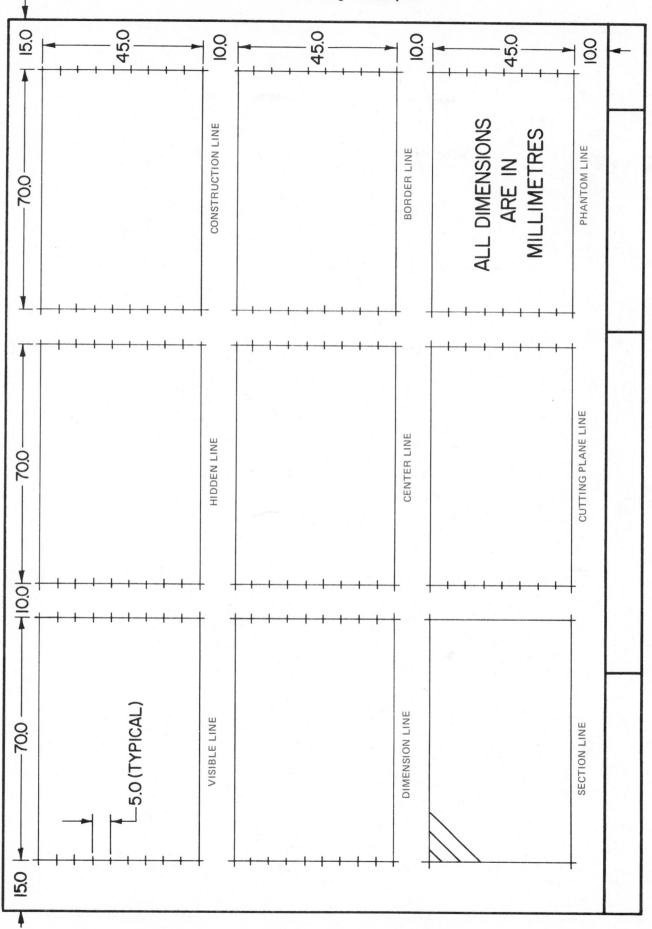

PROBLEM SHEET 4–1. ALPHABET OF LINES. Duplicate this drawing on a separate sheet of paper. Construct the nine different lines called for.

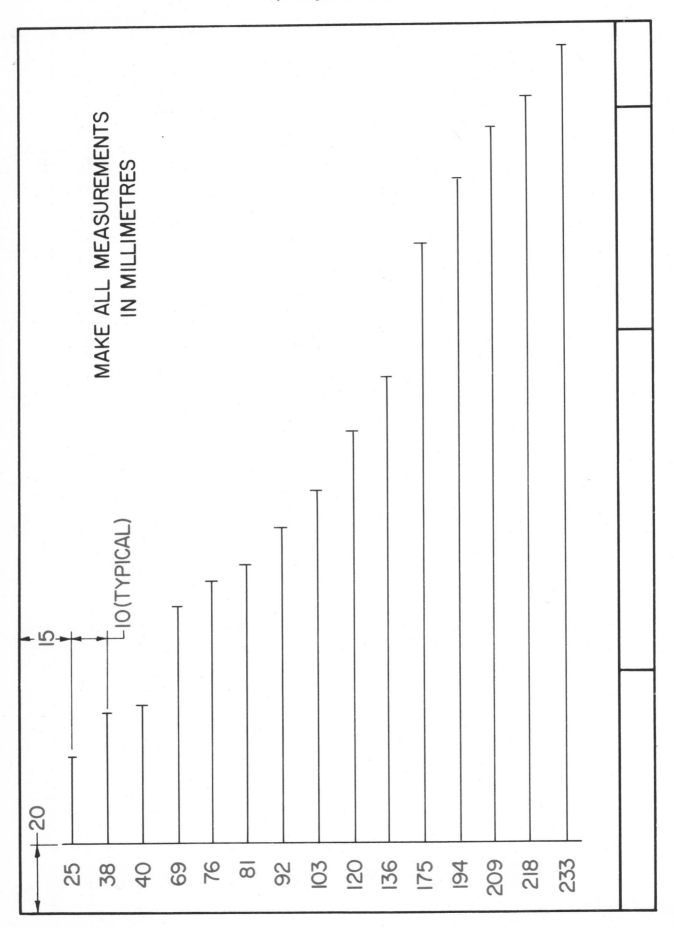

MAKE ALL MEASUREMENTS
IN MILLIMETRES

15

20

10(TYPICAL)

25
38
40
69
76
81
92
103
120
136
175
194
209
218
233

PROBLEM SHEET 4–2. MEASURING PRACTICE. Duplicate this drawing on a separate sheet of paper.

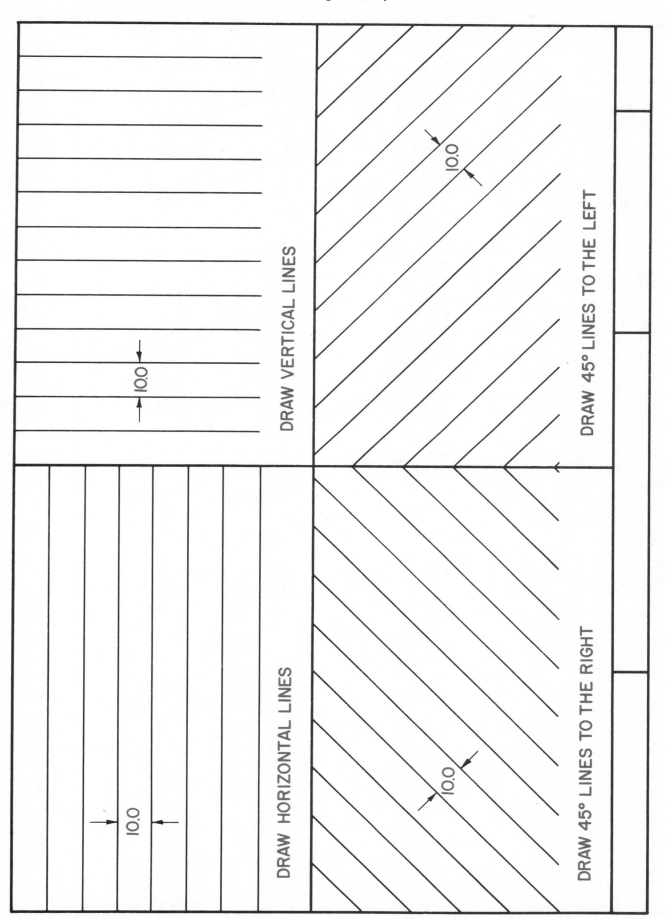

DRAW VERTICAL LINES

10.0

DRAW 45° LINES TO THE LEFT

10.0

DRAW HORIZONTAL LINES

10.0

DRAW 45° LINES TO THE RIGHT

10.0

PROBLEM SHEET 4—3. INSTRUMENT PRACTICE.

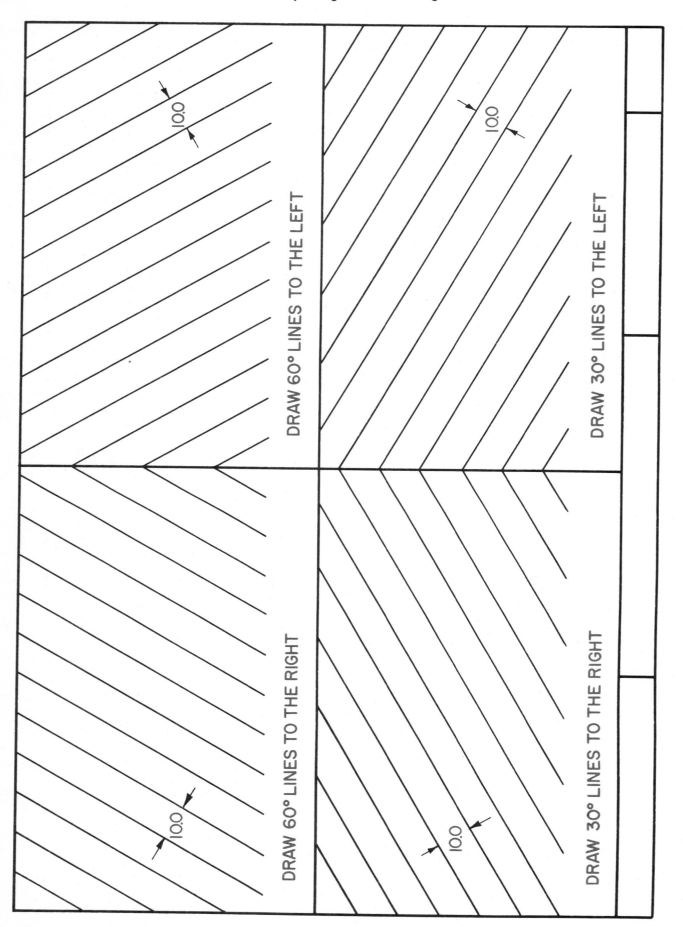

DRAW 60° LINES TO THE LEFT

DRAW 30° LINES TO THE LEFT

DRAW 60° LINES TO THE RIGHT

DRAW 30° LINES TO THE RIGHT

10.0

10.0

10.0

10.0

PROBLEM SHEET 4—4. INSTRUMENT PRACTICE.

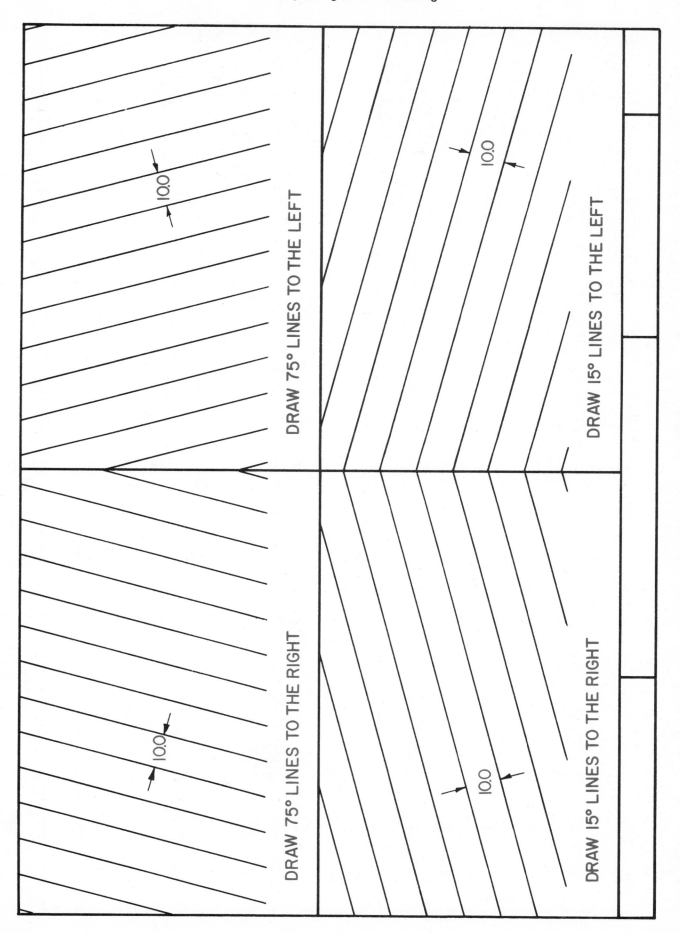

DRAW 75° LINES TO THE LEFT

10.0

DRAW 75° LINES TO THE RIGHT

10.0

DRAW 15° LINES TO THE LEFT

10.0

DRAW 15° LINES TO THE RIGHT

10.0

PROBLEM SHEET 4–5. INSTRUMENT PRACTICE.

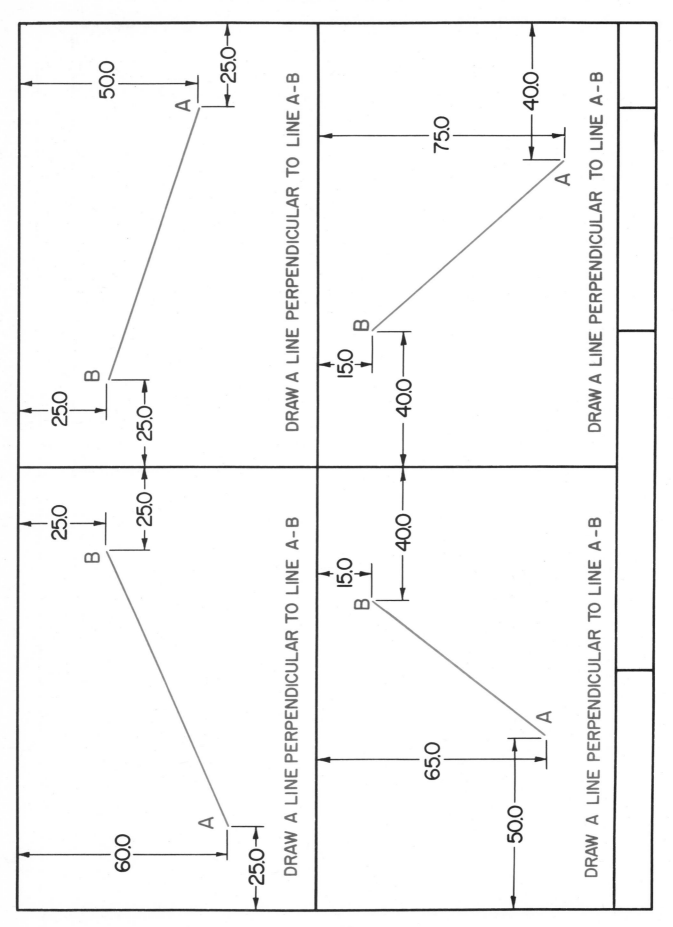

DRAW A LINE PERPENDICULAR TO LINE A-B

DRAW A LINE PERPENDICULAR TO LINE A-B

DRAW A LINE PERPENDICULAR TO LINE A-B

DRAW A LINE PERPENDICULAR TO LINE A-B

PROBLEM SHEET 4–6. INSTRUMENT PRACTICE.

Unit 5
GEOMETRICAL CONSTRUCTION

In your daily activities, you have many encounters with geometry. The various designs of the airplanes that fly overhead and automobiles that pass on the street are based on geometrics. Buildings and bridges utilize squares, rectangles, triangles, circles and arcs in their construction. Every mechanical drawing and project is composed of one or more geometrical shapes. See the examples shown in Fig. 5-1.

It is important that you develop the ability to visualize and draw the basic geometric shapes presented in this Unit. This will aid you in solving drafting problems and help to improve your skill in using drafting instruments.

Fig. 5-1. Wood project designs (novelty boxes) which involve several geometric figures.

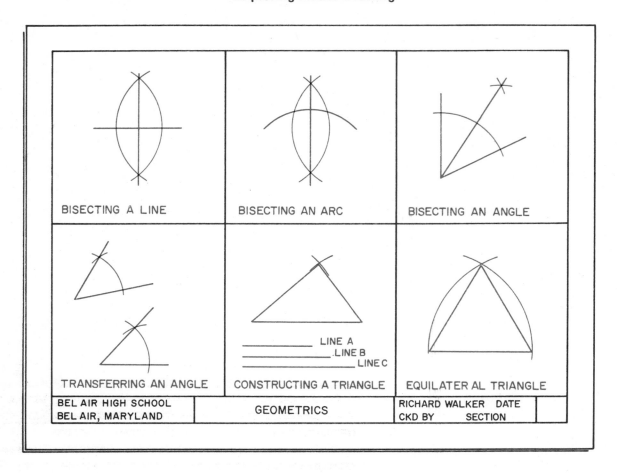

BISECTING A LINE	BISECTING AN ARC	BISECTING AN ANGLE
TRANSFERRING AN ANGLE	CONSTRUCTING A TRIANGLE	EQUILATERAL TRIANGLE

LINE A
LINE B
LINE C

| BEL AIR HIGH SCHOOL BEL AIR, MARYLAND | GEOMETRICS | RICHARD WALKER   DATE CKD BY   SECTION | |

Fig. 5-2. In drawing geometric shapes, several figures may be placed on a single sheet.

DRAWING BASIC GEOMETRIC PROBLEMS

Since size does not enter into the solution of most of the geometric problems presented in this Unit, no dimensions are given. Most of the problems require little space in their solution; therefore, several problems may be included on a single drawing sheet. A suggested sheet layout is given in Fig. 5-2.

The drawing sheets can be made more interesting and attractive if colored pencils are used to draw the lines used to construct the geometric figures.

HOW TO BISECT OR FIND THE MIDDLE OF A LINE

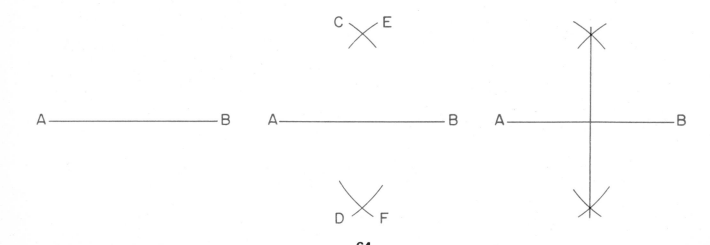

(The bisecting line will be at right angles, 90 deg., to the given line.)

1. Let line A—B be the line to be bisected.
2. Set your compass to a distance larger than one half the length of the line to be bisected. Using this setting as the radius and the end of the line at A as the center point, draw arc C—D. Using the same compass setting but the end of the line at B as the center point, draw arc E—F.
3. Draw a line through the points where the arcs intersect. This line will be at right angles (90 deg. or perpendicular) to and bisect the original line A—B.

HOW TO BISECT AN ARC

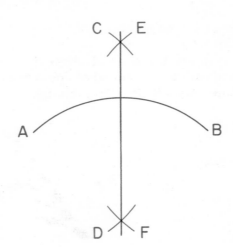

1. An arc or part of a circle is bisected by the same method described in HOW TO BISECT OR FIND THE MIDDLE OF A LINE.

HOW TO BISECT AN ANGLE

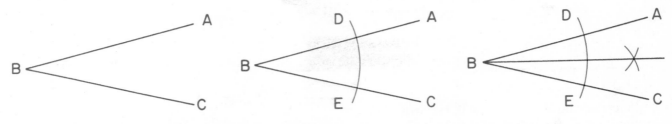

1. Let lines A—B and B—C be the angle to be bisected.
2. With B as the center, draw an arc intersecting the angle at D and E.
3. Using a compass setting greater than one half D—E as centers, draw intersecting arcs. A line through this intersection and B will bisect the angle.

HOW TO TRANSFER OR COPY AN ANGLE

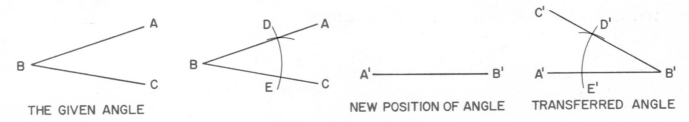

THE GIVEN ANGLE   NEW POSITION OF ANGLE   TRANSFERRED ANGLE

1. Let lines A—B and B—C be the angle to be transferred or copied.
2. Locate the new position of the angle and draw line A'—B'.
3. With B as the center point, draw an arc of any convenient radius on the given angle. This arc intersects the given angle at D and E.
4. Using the same radius and B' as the center, draw arc D'—E'.
5. Set your compass equal to D—E. With point E' as a center and D—E as the radius, strike an arc which intersects the first arc at point D'.
6. Draw a line through the intersecting arcs to complete the transfer of the given angle.

HOW TO CONSTRUCT A TRIANGLE FROM GIVEN LINE LENGTHS

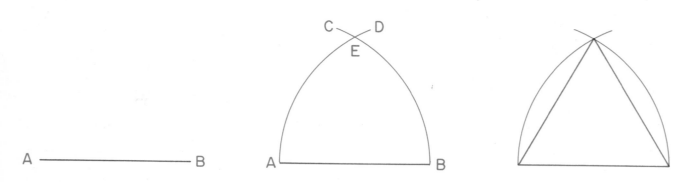

1. Let lines A, B and C be the sides of the required triangle.
2. Draw a line that is equal in length to line A.
3. Set your compass to a length equal to line B. Use one end of line A as a center and strike an arc. Reset your compass to a length equal to line C and with the other end of line A as a center, strike another arc.
4. Connect the ends of line A to the points where the two arcs intersect.

HOW TO CONSTRUCT AN EQUILATERAL TRIANGLE

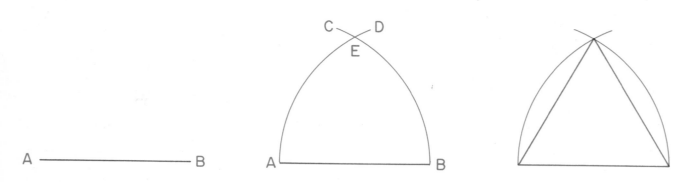

(A triangle having all sides equal in length.)
1. Let line A—B be the length of the sides of the triangle.
2. With A as the center and with the compass setting equal to the length of line A—B, strike the arc B—C. Using the same compass setting but with B as the center, strike the arc A—D. These arcs intersect at E.
3. Complete the triangle by connecting A to E and B to E.

HOW TO DRAW A SQUARE WITH THE DIAGONAL GIVEN

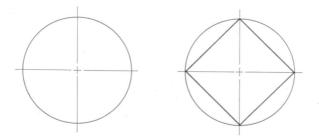

1. Draw a circle with a diameter equal to the length of the diagonal.
2. Connect the points where the center lines intersect the circle.

HOW TO DRAW A SQUARE WITH THE SIDE GIVEN

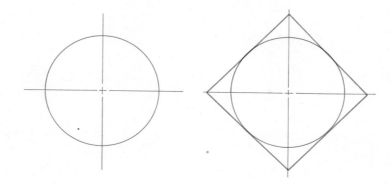

1. Draw a circle with a diameter equal to the length of the side.
2. Draw tangents 45 deg. to the center line.

HOW TO CONSTRUCT A PENTAGON OR FIVE POINT STAR

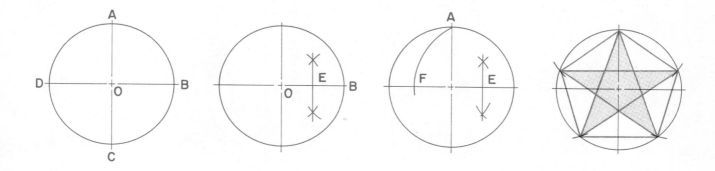

1. Draw a circle. Let A—C and B—D be the center lines and O be the point where the center lines intersect.
2. Bisect the line (radius) O—B. This will locate point E.
3. With E as the center and with the compass set to the radius E—A, strike the arc A—F.
4. The distance A—F is one-fifth the circumference of the circle. Set your compass or dividers to this distance and circumscribe the circle, starting at A. Connect these points as a pentagon or as a five point star.

HOW TO CONSTRUCT A HEXAGON (FIRST METHOD)

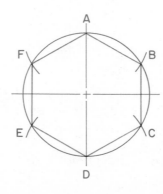

1. Draw a circle.
2. With a compass setting equal to the radius of the drawn circle, start a point A and circumscribe the circle locating points B, C, D, E, F.
3. Connect the points with a straightedge to complete the hexagon.

HOW TO CONSTRUCT A HEXAGON (SECOND METHOD)

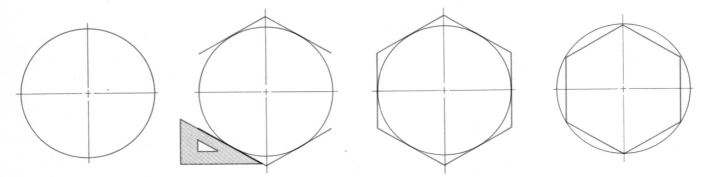

1. Draw a circle.
2. Use the 30-60 deg. triangle and draw construction lines at 60 deg. to the vertical center line and tangent to the circle.
3. Draw two vertical construction lines tangent to the circle. Fill in the construction lines with visible object lines to complete the hexagon.
4. The same basic technique can be used to draw (inscribe) a hexagon inside the circle.

HOW TO DRAW AN OCTAGON USING A CIRCLE

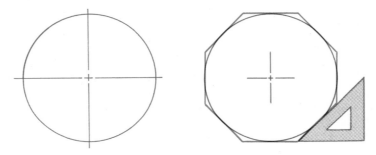
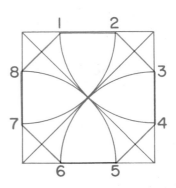

1. Draw a circle with a diameter equal to the distance across the flats of the desired octagon.
2. Draw the vertical and horizontal lines tangent to the circle. Use construction lines.
3. Complete the octagon by drawing the 45 deg. angle lines tangent to the circle. Fill in the construction lines with visible object lines.

HOW TO DRAW AN OCTAGON USING A SQUARE

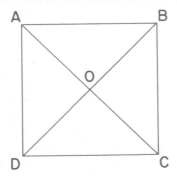
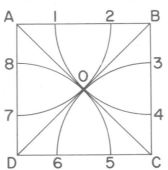

1. Draw a square with the sides equal in length to the distance across the flats of the required octagon. Draw the diagonals A—C and B—D. The diagonals intersect at O.
2. Set your compass to radius A—O and, with corners A, B, C and D as centers, draw arcs that intersect the square at points 1, 2, 3, 4, 5, 6, 7 and 8.
3. Complete the octagon by connecting point 1 to 2, 2 to 3, 3 to 4, 4 to 5, 5 to 6, 6 to 7, 7 to 8 and 8 to 1 with object lines.

HOW TO DRAW AN ARC TANGENT TO TWO LINES AT A RIGHT ANGLE

 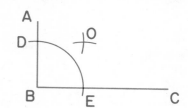 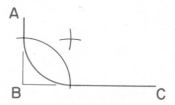

1. Let A—B and B—C be the lines that form the right angle.
2. Set your compass to the radius of the required arc and using B as the center, strike arc D—E. With D and E as centers and with the same compass setting, draw the arcs that intersect at O.
3. With O as the center and with the compass at the same setting, draw the required arc. It will be tangent to Lines A—B and B—C at points D and E.

HOW TO DRAW AN ARC TANGENT TO TWO STRAIGHT LINES

 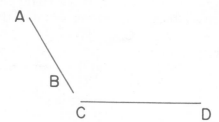 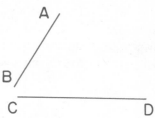

 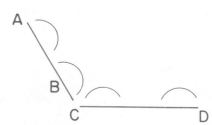 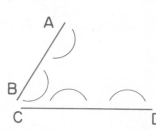

 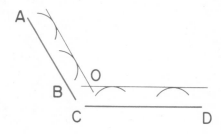 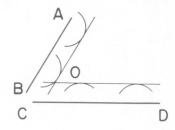

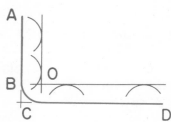 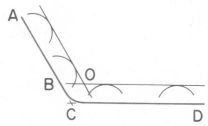 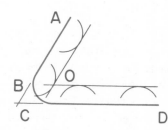

1. Let lines A—B and C—D be the two straight lines.
2. Set your compass to the radius of the arc to be drawn tangent to the two straight lines and with points near the ends of the lines A—B and C—D as centers, strike two arcs on each line.
3. Draw straight construction lines tangent to the arcs.
4. The point where the two lines intersect (O) is the center for drawing the required arcs.

HOW TO DRAW AN ARC TANGENT WITH A STRAIGHT LINE AND A GIVEN ARC

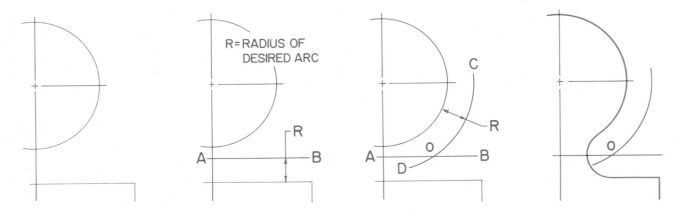

1. Draw the given arc and the straight line in proper relation to one another. Use construction lines.
2. Draw line A—B a distance equal to the radius of the desired arc from and parallel to the straight line.
3. Draw arc C—D by setting your compass to a radius equal to the radius of the given arc plus the radius of the desired arc. This arc intersects with line A—B at point O.
4. Using point O as the center, and with the compass set to the radius of the desired arc, draw the required arc. It will be tangent to the given arc and to the straight line. Fill in construction lines with object lines.

HOW TO DRAW TANGENT ARCS

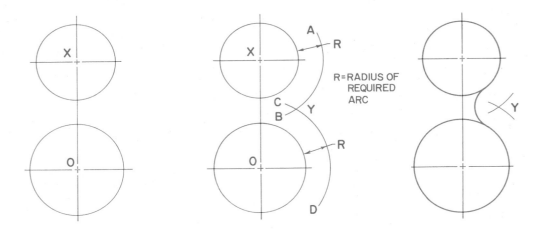

1. Draw the two arcs in proper relation to each other that require the tangent arc to join them. Let O and X be the centers of these arcs.
2. Set the compass to a distance equal to the radius of the given circle plus the radius of the required arc. Using X as the center, draw arc A—B. Reset the compass to a distance equal to the radius of the second given circle plus the radius of the required arc. Using O as the center, draw arc C—D. These arcs intersect at Y.
3. Set the compass to the radius of the required arc. With Y as the center, draw the desired arc. This arc will be tangent to the given arcs.

HOW TO DIVIDE A LINE INTO A GIVEN NUMBER OF EQUAL DIVISIONS

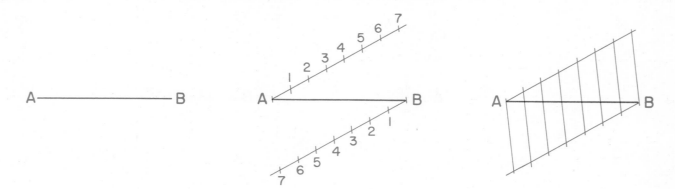

1. PROBLEM: Divide line A—B into seven equal divisions.
2. Project construction lines from A and B that are parallel to one another.
3. Open your compass or dividers to a suitable setting (approximately 5 mm) and, starting at A, step off seven spaces on the projected line. Starting at B, step off seven spaces on the other projected line.
4. Connect the points with construction lines as shown. The points of division are where these lines pass through the given line A—B.

HOW TO DRAW AN ELLIPSE USING CONCENTRIC CIRCLES

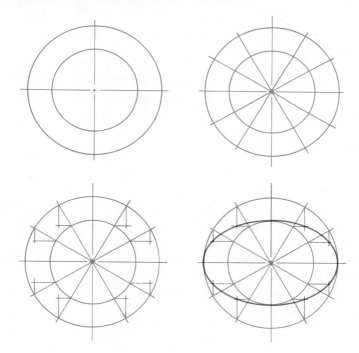

1. Draw two concentric circles. The diameter of the large circle is equal to the length of the large axis of the desired ellipse. The diameter of the small circle is equal to the length of the small axis of the desired ellipse.
2. Divide the two circles into twelve equal parts. Use your 30-60 deg. triangle.
3. Draw horizontal lines from the points where the dividing lines intersect the small circle. Vertical lines are drawn from the points where the dividing lines intersect the large circle.
4. Connect the points where the vertical and horizontal lines intersect with a French curve.

HOW TO DRAW AN ELLIPSE USING THE PARALLELOGRAM METHOD

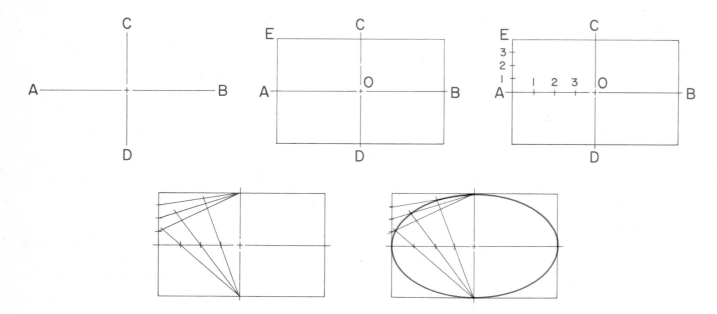

This method is satisfactory for drawing large ellipses.

1. Let line A—B be the major axis and line C—D be the minor axis of the required ellipse.
2. Construct a rectangle with sides equal in length and parallel to the axes.
3. Divide A—O and A—E into the same number of equal parts.
4. From C draw a line to point 1 on line A—E. Draw a line from D through point 1 on line A—C. The point of intersection of these two lines will establish the first point of the ellipse. The remaining points in this section are completed as are similar points in the other three sections (quadrants).
5. Connect the points with a French curve to complete the ellipse.

OUTSIDE ACTIVITIES

1. Prepare a bulletin board display that illustrates the use of geometric shapes in buildings and bridges.
2. Develop a list of everyday items that make use of geometric shapes. For example: Nut and bolt heads are round, square and hexagonal.
3. Secure a photo or a drawing of a modern airplane. Place a sheet of tracing vellum over it and sketch in the various geometric shapes used in its design.
4. Do the same using a photo or drawing of a late model automobile.

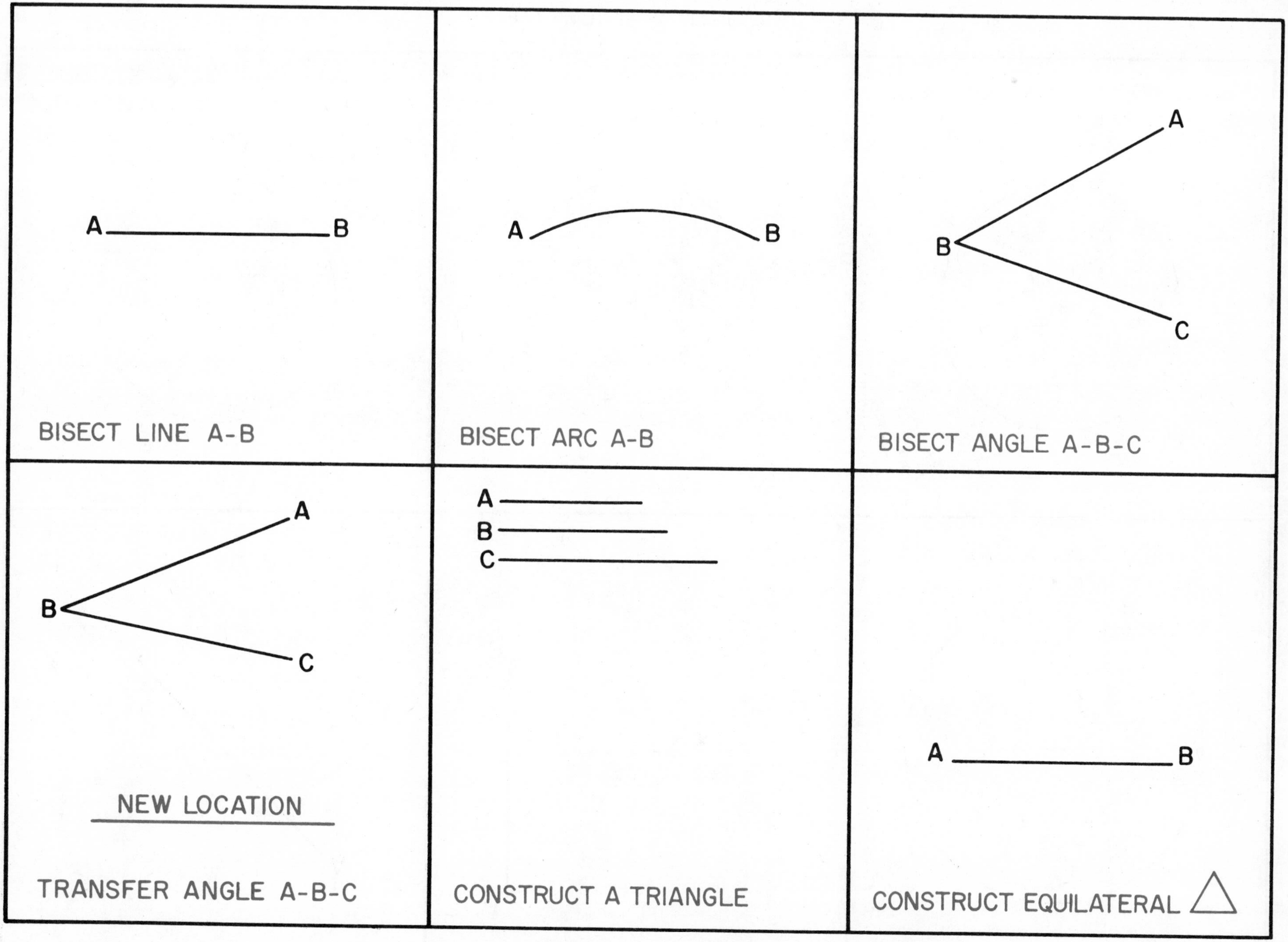

BISECT ANGLE A-B-C

BISECT ARC A-B

BISECT LINE A-B

A

B

C

A

B

A

B

CONSTRUCT EQUILATERAL △

CONSTRUCT A TRIANGLE

A

B

C

A

B

TRANSFER ANGLE A-B-C

NEW LOCATION

A

B

C

PROBLEM SHEET 5—1. GEOMETRICS.

CONSTRUCT A PENTAGON

INSCRIBE A HEXAGON

CONSTRUCT A SQUARE GIVEN
SIDE A-B

A ———————— B

CONSTRUCT A HEXAGON USING
THE 30°-60° TRIANGLE

CONSTRUCT A SQUARE GIVEN
DIAGONAL A-B

A
B

CONSTRUCT A HEXAGON USING
THE COMPASS

PROBLEM SHEET 5—2. GEOMETRICS.

DIVIDE LINE A-B INTO NINE (9) EQUAL PARTS

A ────────── B

DRAW AN ARC TANGENT TO LINES A-B AND B-C

CONSTRUCT AN OCTAGON

DRAW AN ARC TANGENT TO LINES A-B AND B-C

CONSTRUCT AN OCTAGON

DRAW AN ARC TANGENT TO LINES A-B AND B-C

PROBLEM SHEET 5—3. GEOMETRICS.

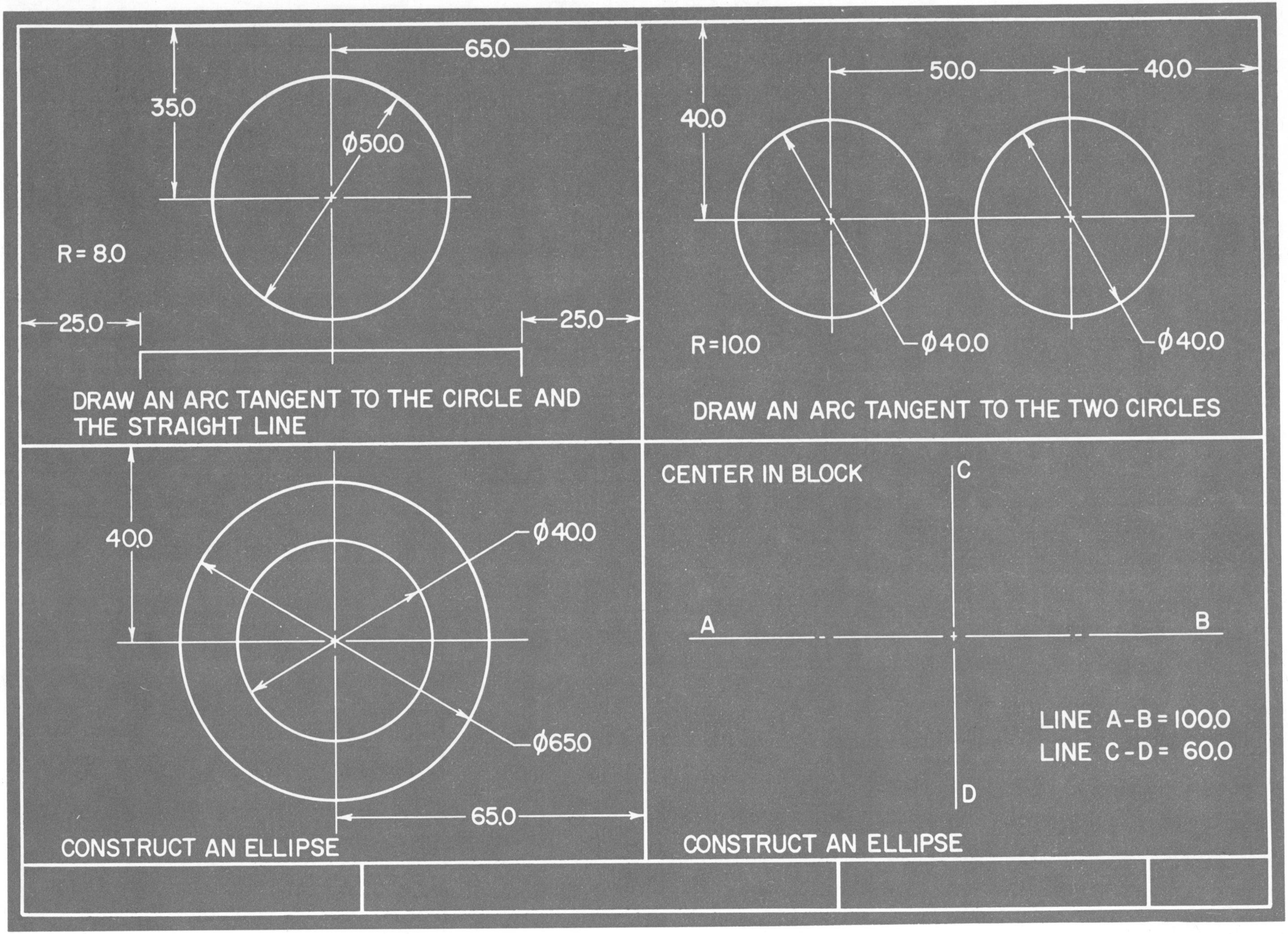

Φ40.0

40.0

50.0

Φ40.0

40.0

R=10.0

DRAW AN ARC TANGENT TO THE TWO CIRCLES

CENTER IN BLOCK

C

A

B

D

LINE A-B = 100.0
LINE C-D = 60.0

CONSTRUCT AN ELLIPSE

65.0

25.0

Φ50.0

35.0

R= 8.0

25.0

DRAW AN ARC TANGENT TO THE CIRCLE AND
THE STRAIGHT LINE

Φ40.0

Φ65.0

65.0

400

CONSTRUCT AN ELLIPSE

PROBLEM SHEET 5—4. GEOMETRICS.

AIRCRAFT INSIGNIA OF THE WORLD

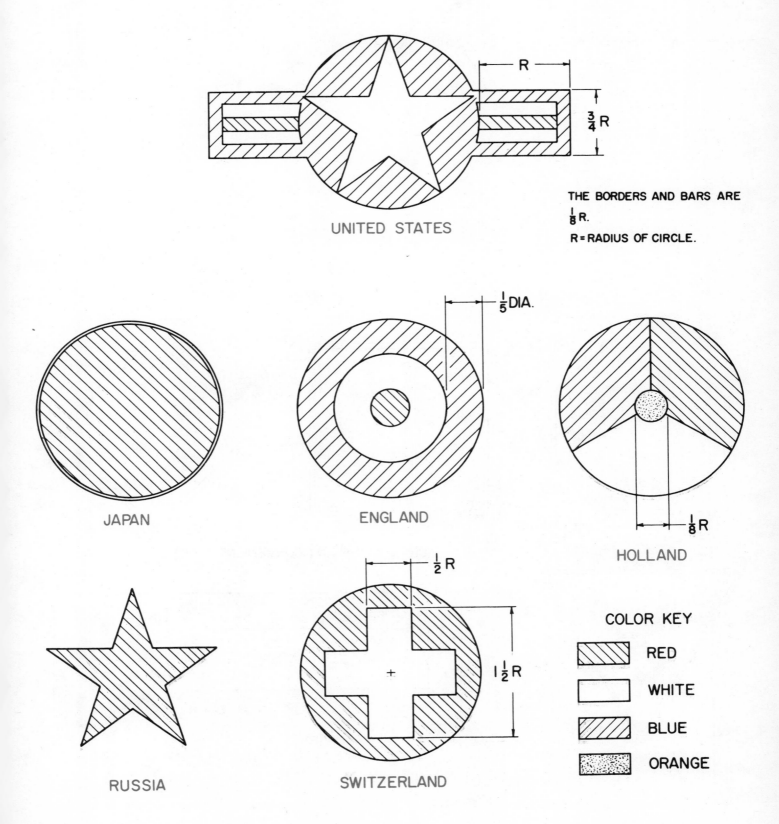

UNITED STATES

THE BORDERS AND BARS ARE $\frac{1}{8}$ R.

R = RADIUS OF CIRCLE.

JAPAN

ENGLAND

$\frac{1}{5}$ DIA.

HOLLAND

$\frac{1}{8}$ R

RUSSIA

SWITZERLAND

$\frac{1}{2}$ R

$1\frac{1}{2}$ R

COLOR KEY

RED

WHITE

BLUE

ORANGE

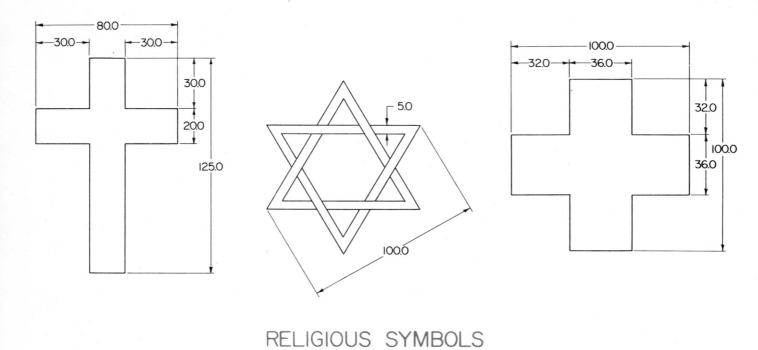

RELIGIOUS SYMBOLS

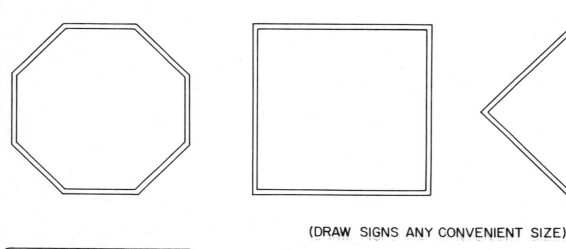

(DRAW SIGNS ANY CONVENIENT SIZE)

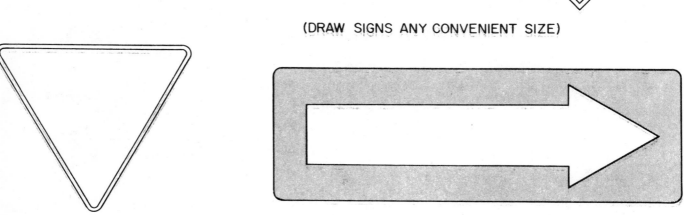

TRAFFIC WARNING SIGNS
HOW IS EACH SIGN USED?

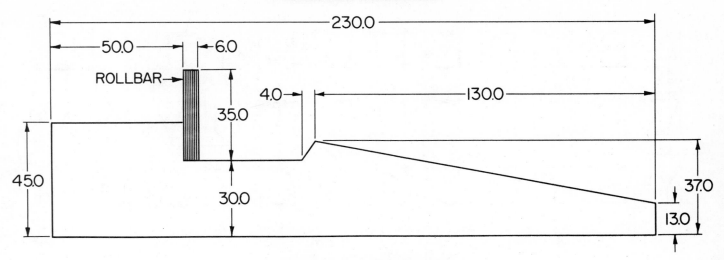

DESIGN PROBLEM 1

DEVELOP DISTINCTIVE RACING STRIPES FOR THIS
FORMULA "V" BODY

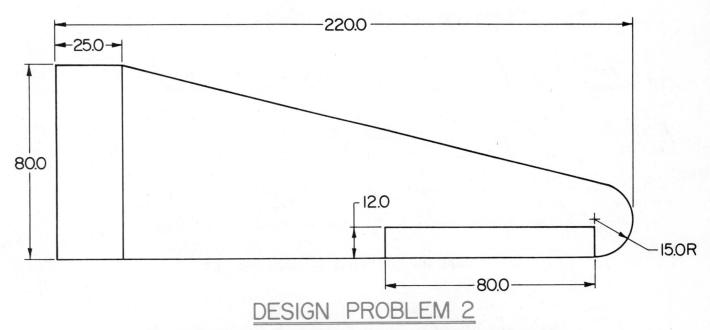

DESIGN PROBLEM 2

DESIGN A UNIQUE COLOR SCHEME FOR THE AMERICAN
AIRPLANE TO BE USED IN THE WORLD AEROBATIC
CHAMPIONSHIP. USE GEOMETRIC FIGURES.

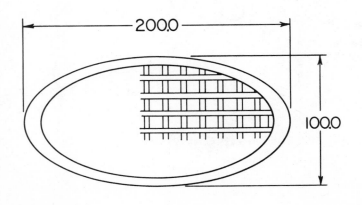

DESIGN PROBLEM 3

DESIGN A GRILLE FOR THIS SPORT CAR
FRONT.

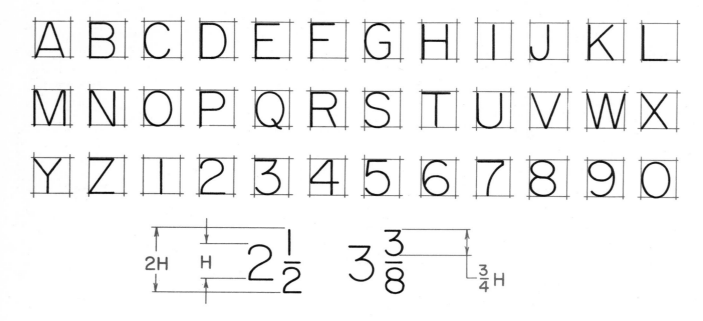

VERTICAL SINGLE STROKE GOTHIC ALPHABET

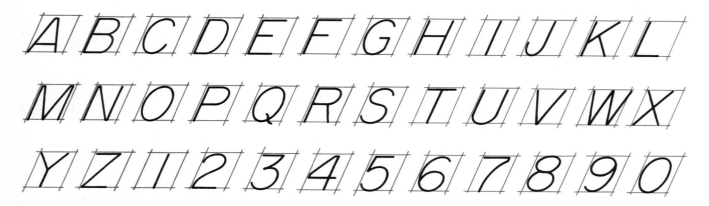

INCLINED SINGLE STROKE GOTHIC ALPHABET

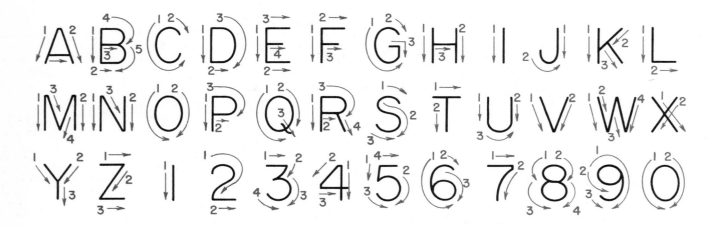

RECOMMENDED SEQUENCE FOR MAKING SINGLE STROKE GOTHIC ALPHABET

Fig. 6-1. The Single Stroke Gothic Alphabet.

Unit 6
LETTERING

LETTERING is used on drawings to give dimensions and other pertinent information needed to fully describe the item. The lettering must be neat and legible if it is to be easily read and understood.

A drawing will be improved by good lettering. A good drawing will look sloppy and unprofessional if the lettering is poorly done.

SINGLE STROKE GOTHIC ALPHABET

The American National Standards Institute (originally ASA) recommends that SINGLE STROKE GOTHIC ALPHABET, Fig. 6-1, be the accepted lettering standard because it can be drawn rapidly and is highly legible. It is called single stroke lettering not because each letter is made with a single stroke of the pencil (most letters require several strokes to complete), but because each line is only as wide as the point of the pencil or pen.

Single stroke lettering may be vertical or inclined, as there is no definite rule stating that it should be one or the other. However, mixing them on a drawing should be avoided, Fig. 6-2.

YOU can do first class lettering if you learn the basic shapes of the letters, the proper stroke sequence and the recommended spacing between letters and words. You must also practice regularly.

ONLY ONE FORM OF LETTERING SHOULD
APPEAR ON A DRAWING.

AVOID COMbINING SEVERAL fORMS Of
LETTERING.

Fig. 6-2. Avoid mixing several forms of lettering on a drawing.

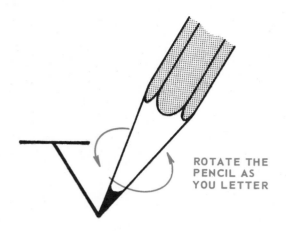

Fig. 6-3. Rotate the pencil point as you letter to keep the point sharp and the lettering uniform in weight.

LETTERING WITH A PENCIL

A H or 2H pencil is used by most drafters for lettering. Sharpen the pencil to a sharp conical point, and rotate it as you letter, Fig. 6-3, to keep the point sharp and the letters uniform in weight and line width. Resharpen the pencil when the lines become wide and "fuzzy."

GUIDE LINES

Good lettering requires the use of GUIDE LINES, Fig. 6-4. Guide lines are very fine lines made with a

DEVALUATION

SPACED BY MEASURING

DEVALUATION

SPACED VISUALLY

Fig. 6-7. Letter spacing is judged by eye rather than by measuring.

"needle sharp" 4H or 6H pencil. Guide lines should be drawn so lightly they will not show up on a print made from the drawing. VERTICAL GUIDE LINES, Fig. 6-5, may be used to assure that the letters will be vertical. Use INCLINED GUIDE LINES, Fig. 6-6, drawn at 67 1/2 deg. to the horizontal line, where inclined lettering is to be used.

SPACING

In lettering, proper spacing of the letters is important. There is no hard and fast rule which indicates how far apart the letters should be spaced. The letters should be placed so spaces between the letters appear to be about the same. Adjacent letters with straight lines require more space than curved letters. Letter spacing is judged by eye rather than by measuring, Fig. 6-7.

ALWAYS USE GUIDE LINES WHEN
LETTERING. THEY ARE NEEDED.

Fig. 6-4. Guide lines must be used when lettering to keep the letters uniform in height.

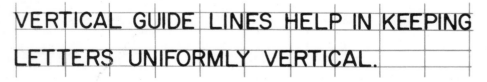

Fig. 6-5. Vertical guide lines.

Fig. 6-6. Inclined guide lines.

WORDS AND LETTERS MUST BE CLEARLY SEPARATED. SPACING BETWEEN WORDS AND SENTENCES IS EQUAL TO THE HEIGHT OF THE LETTER USED.

Fig. 6-8. Spacing between words and sentences.

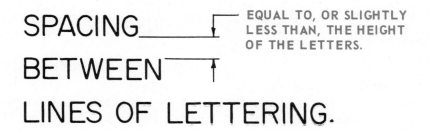

SPACING — EQUAL TO, OR SLIGHTLY LESS THAN, THE HEIGHT OF THE LETTERS.

BETWEEN

LINES OF LETTERING.

Fig. 6-9. Spacing between lines of lettering.

Spacing between words and between sentences is another matter. The spacing between words and sentences should be equal to the height of the letters, Fig. 6-8.

Spacing between lines of lettering should be equal to, or slightly less than, the height of the letters, as shown in Fig. 6-9. Proper spacing — between words, sentences and lines — improves the appearance of the lettering and makes it easier to read.

LETTER HEIGHT

Letters 3.0 or 4.0 mm high are satisfactory on A4 size sheets. Letter sizes should be increased to 5.0 mm on A3 size sheets.

LETTERING AIDS AND LETTERING DEVICES

Various lettering aids and devices are available. The LETTERING TRIANGLE, Fig. 6-10, and the

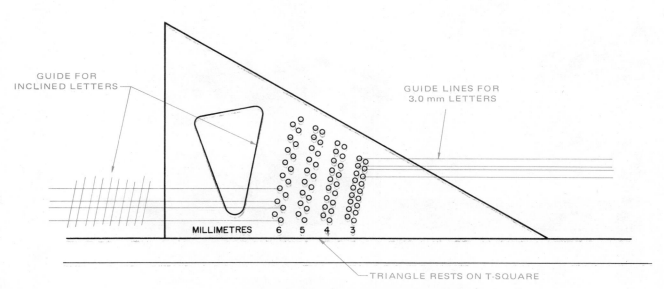

Fig. 6-10. The Braddock lettering triangle.

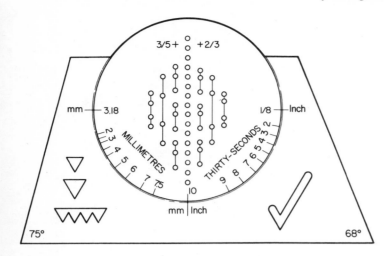

Fig. 6-11. The Ames lettering instrument.

AMES LETTERING INSTRUMENT, Fig. 6-11, are devices that may be used as aids for drawing guide lines. The numbers engraved below each series of holes in the triangle indicate the height of the letters in millimetres. For example, the series marked 4 indicates the guide lines will be 4.0 millimetres apart. The numbers on the disk of the Ames lettering instrument are rotated until they are even with a line

engraved on the base of the tool. The numbers also indicate the spacing of the guide lines in millimetres or thirty-seconds of an inch.

MECHANICAL LETTERING DEVICES

Hand lettering is expensive because it requires so much time. For this reason, modern industry utilizes many mechanical devices to save time, and to improve the legibility of the letters.

Many drafting and engineering offices are making use of typewriters, Fig. 6-12, to put information on drawings. This releases highly skilled drafters for other work.

With the advent of microfilm, it became necessary for lettering to be highly legible (the drawings are reduced and enlarged photographically). Good lettering can be done easily and rapidly with mechanical lettering devices. See Fig. 6-13. This type equipment is available for a great variety of lettering styles, alphabets and symbols in a wide range of sizes and thicknesses. The lettering equipment shown in Fig. 6-14 is used for making signs and posters.

Fig. 6-12. A special typewriter may be used to put information on drawings.
(Mechanical Enterprises, Inc.)

84

Fig. 6-13. LEROY lettering equipment produces lettering mechanically. Many styles of lettering are available.

TEST YOUR KNOWLEDGE - UNIT 6

(Write answers on a separate sheet of paper.)
1. Why is lettering needed on drawings?
2. The single stroke gothic letter is recommended because _____ .
3. The single stroke lettering used on a drawing may be _____ or _____ because there is no definite rule stating that it has to be one or the other.
4. The _____ or _____ pencil is usually used for lettering.
5. Why are guide lines used when lettering?
6. Name three types of lettering aids and mechanical lettering devices.
7. Why does industry use mechanical lettering devices?

OUTSIDE ACTIVITIES

1. Using signs and posters from your school's bulletin board, make a display of different lettering examples. Point out good spacing and poor spacing. Check to see if lettering styles are mixed in each example.
2. Research the variety of lettering styles, alphabets, and symbols available with mechanical lettering devices.
3. Letter the sentence "Good lettering technique requires practice and concentration " by hand, using Vertical Single Stroke Gothic. Letter the same sentence using a mechanical lettering device such as the LEROY lettering equipment. Time yourself and report to the class which is faster and which is easier to read.

Fig. 6-14. The Wrico Sign-Maker utilizes a plastic stencil to make characters. This model is used for making large lettering for signs and posters. (Wood-Regen Instrument Co., Inc.)

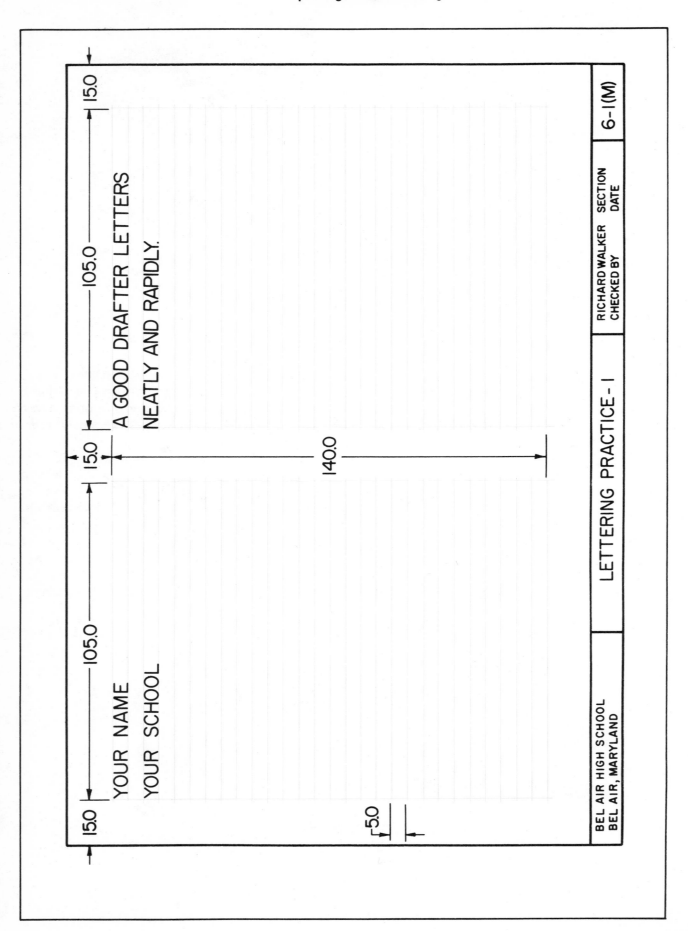

PROBLEM SHEET 6—1. LETTERING PRACTICE.

THE QUICK RED FOX JUMPED OVER
THE LAZY BROWN DOG.

1 2 3 4 5 6 7 8 9 0

LETTERING PRACTICE - 2

6-2(M)

PROBLEM SHEET 6–2. LETTERING PRACTICE.

"ONE SMALL STEP FOR A MAN,
ONE GIANT LEAP FOR MANKIND"

"HERE MEN FROM THE PLANET
EARTH FIRST SET FOOT UPON
THE MOON JULY 1969, A.D. WE
CAME IN PEACE FOR ALL MAN-
KIND."

6-3(M)

LETTERING PRACTICE - 3

PROBLEM SHEET 6—3. LETTERING PRACTICE.

THERE IS NEVER ENOUGH TIME

TO DO A JOB PROPERLY, BUT

IT SEEMS THERE IS ALWAYS

PLENTY OF TIME TO DO THE

JOB OVER.

1234567890

6-4(M)

LETTERING PRACTICE-4

PROBLEM SHEET 6—4. LETTERING PRACTICE.

SELECT A FAVORITE SAYING OR
QUOTATION AND LETTER IT IN
3.0 mm VERTICAL LETTERS.

6-5(M)

LETTERING PRACTICE-5

PROBLEM SHEET 6–5. LETTERING PRACTICE.

SELECT A FAVORITE SAYING OR
QUOTATION AND LETTER IT IN
3.0 mm INCLINED LETTERS.

6-6(M)

LETTERING PRACTICE-6

PROBLEM SHEET 6–6. LETTERING PRACTICE.

Unit 7
MULTIVIEW DRAWINGS

When a drawing is made with the aid of instruments, it is called a MECHANICAL DRAWING. Straight lines are drawn using a T-square and triangle or drafting machine, while circles, arcs and curves are drawn with a compass and French curve.

Almost all drawings used by industry are made using instruments and are in the form of MULTIVIEW DRAWINGS. That is, more than one view is required to give a shape description of the object being drawn. In developing the needed views, the object is normally viewed from six directions, as shown in Fig. 7-1.

The various directions of sight will give us the FRONT, TOP, RIGHT SIDE, LEFT SIDE, REAR and BOTTOM VIEWS, Fig. 7-2. To obtain the views,

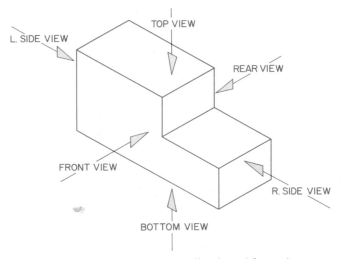

Fig. 7-1. An object is normally viewed from six different directions.

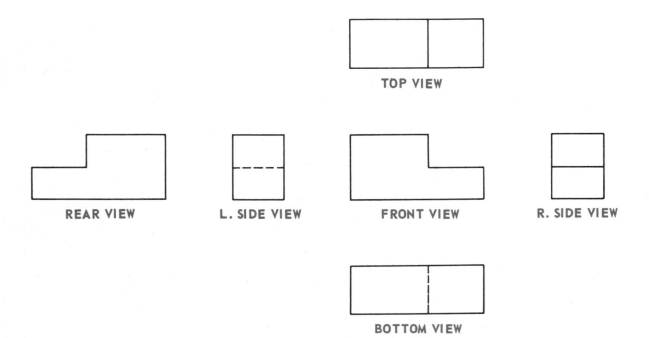

Fig. 7-2. The six directions of sight give these views.

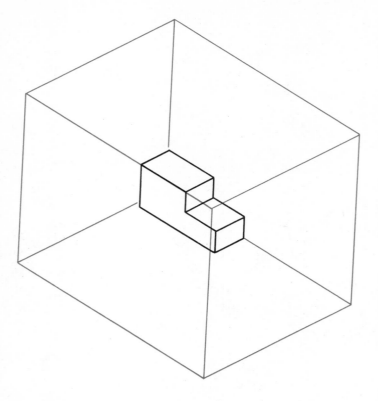

Fig. 7-3. The object is enclosed in a hinged glass box.

think of the object as being enclosed in a hinged glass box, Fig. 7-3.

The method of developing multiview drawings is called ORTHOGRAPHIC PROJECTION. It permits three dimensional objects to be drawn on a flat sheet of paper having only two dimensions. It reveals the width, depth and height of the object.

Orthographic projection forms the basis for engineering drawing. Two methods of projection are used, Fig. 7-4. THIRD ANGLE PROJECTION is preferred by the United States and several other nations. FIRST ANGLE PROJECTION is used in Europe.

With third angle projection, the object IS DRAWN AS VIEWED IN A GLASS BOX. See Fig. 7-5. That is, the views are projected to the six sides of the box. The projected views are drawn as shown when the glass box is opened out.

With first angle projection, the object IS DRAWN AS IF THE OBJECT WERE SET ON THE DRAWING SURFACE. See Fig. 7-6.

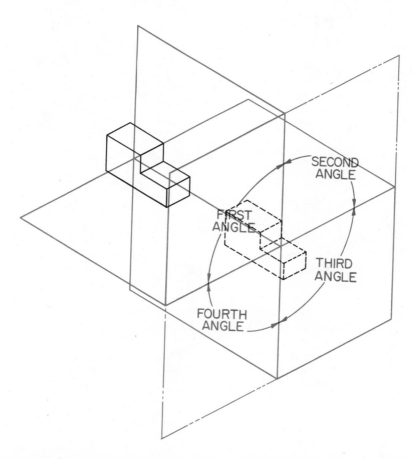

Fig. 7-4. Two methods of projection generally are employed: first angle projection in Europe; third angle projection in the United States and many other countries.

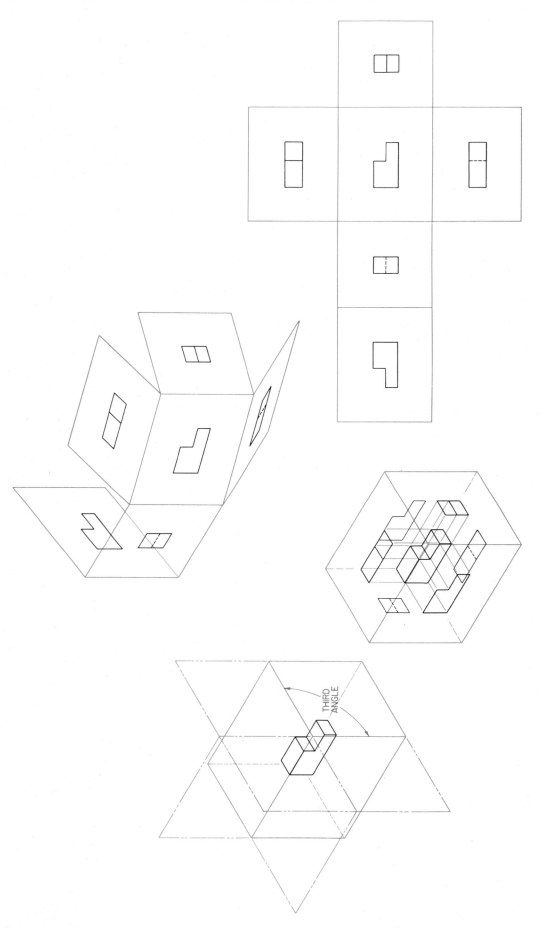

Fig. 7-5. A pictorial explanation of third angle projection.

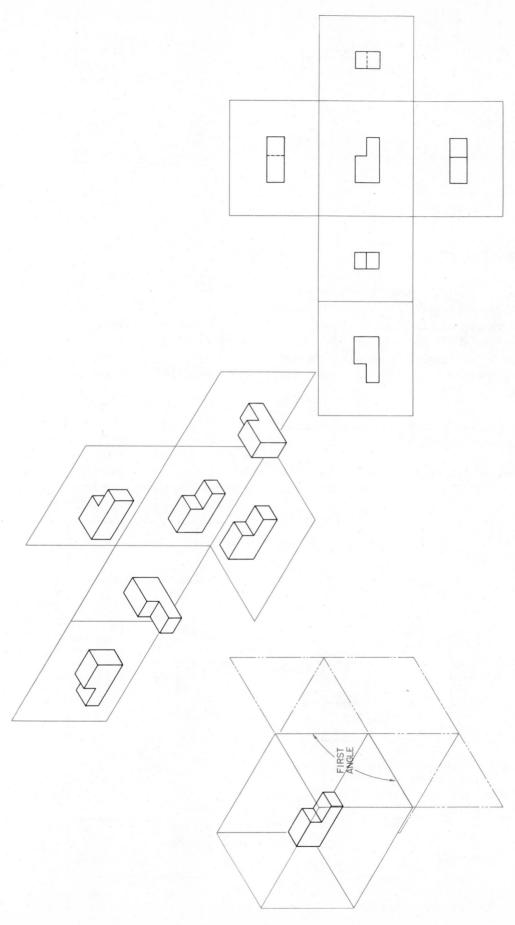

Fig. 7-6. A pictorial explanation of first angle projection.

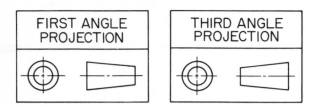

Fig. 7-7. Appropriate ISO symbol is placed on drawing sheet to show which method of projection is used on that sheet.

The ISO symbols shown in Fig. 7-7 are used to indicate the projection system employed on a given drawing.

SELECTING VIEWS TO BE USED

As you can see by Figs. 7-5 and 7-6, we can draw at least six views of the object. This does not mean that all of these views must be used, or are needed. Only those views needed to give a shape description of the object should be drawn. Any view that repeats the same shape description as another view need not be used, Fig. 7-8.

In most instances, two or three views are sufficient to show the shape of an object.

Views showing a large number of hidden lines should be used only if absolutely necessary. Too

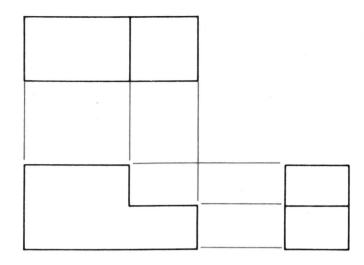

Fig. 7-9. Transferring points from view to view.

many hidden lines tend to make the drawing confusing. Use another view or a sectional view.

TRANSFERRING POINTS

Each view will show a minimum of two dimensions. Any two views of an object will have at least one dimension in common. Time can be saved if a dimension from one view is projected to the other view instead of measuring, Fig. 7-9. Construction lines are used when transferring points.

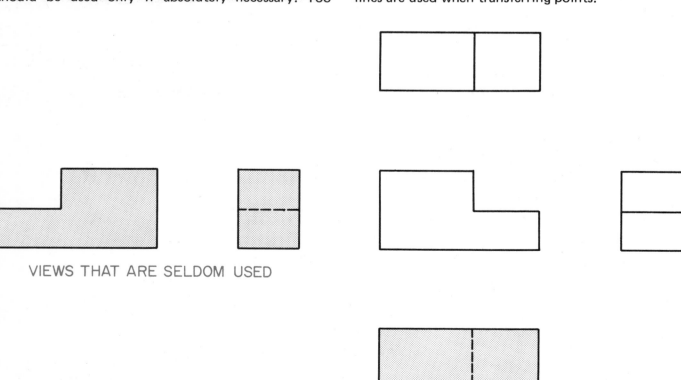

VIEWS THAT ARE SELDOM USED

Fig. 7-8. Not all views are needed. Eliminate any view that repeats the same shape description as another view.

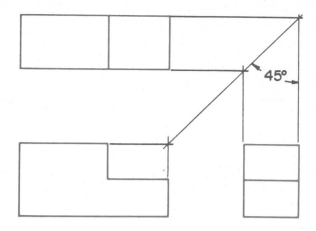 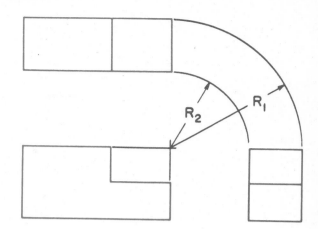

Fig. 7-10. Two accepted methods used to transfer the depth of the top view to the side view.

Additional time can be saved in transferring the depth of the top view to the side view. Two methods of projection are shown in Fig. 7-10. Projection provides for greater accuracy in the alignment of the views, and it is faster than measuring each view separately with a scale or dividers. With this method, a 45 deg. miter line is used to transfer the depth measurement. At right, a compass is used to project these measurements.

HOW TO CENTER DRAWING
ON SHEET

A drawing looks more professional if the views are evenly spaced and centered on the drawing sheet. Centering the views on the sheet is not difficult.

The following procedure is recommended:
1. Examine the object that is to be drawn. Observe its dimensions — width, depth and height, Fig. 7-11. Determine the position in which the object is to be drawn.

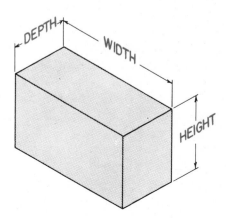

Fig. 7-11. How an object is described.

2. Allow 25 mm between the front and top views and between the front and side views. Measure the working area of your sheet after the title block and border have been drawn. It should measure 180 mm by 260 mm if you used an 8 1/2 in. by 11 in. drawing sheet.

3. To locate the front view, add the width of the front view to the depth of the right side view, plus the 25 mm space between the views. Subtract this total from the horizontal width of the working surface. Divide this answer by two. This will be the starting point for laying out the sheet horizontally. Measure in the resulting answer from the left border line and draw a vertical construction line. With this construction line as the reference point, measure over a distance equal to the width of the front view and draw another vertical construction line through this point, Fig. 7-12.

4. The same procedure is used to center the views vertically. However, the height of the front view and the depth of the top view are used. A 25 mm space will separate the views. Add these distances, and subtract the answer from the vertical working distance of the paper. Divide the resulting answer by two. This distance is measured vertically from the top of the title block. From this starting point, measure up the height of the front view, the 25 mm that separates the views and the depth of the top view. Draw construction lines through these points, Fig. 7-13.

5. Use either the 45 deg. angle method or the radius method, to transfer the depth of the top view to the right side view of the object, Fig. 7-14.

6. Draw in the right side view. Use construction lines.

7. Complete the drawing by going over the proper construction lines. Use the correct weight and type of line (object line, hidden object line, center lines, etc.), Fig. 7-15.

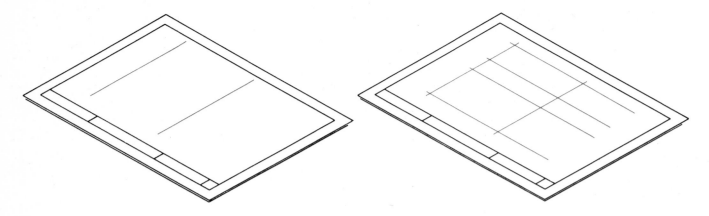

Fig. 7-12. Left. The first step in locating the front and top views on the drawing sheet. Fig. 7-13. Right. The front and top views are blocked in.

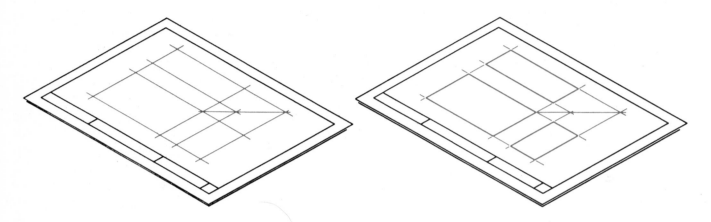

Fig. 7-14. Left. Projecting blocking in the right side view. Fig. 7-15. Right. Draw in the object lines to complete the drawing. Construction lines may be erased.

TEST YOUR KNOWLEDGE - UNIT 7

(Write answers on a separate sheet of paper.)

1. A drawing is said to be a MECHANICAL DRAW-ING when it was drawn using _____.
2. A drawing that uses two or more views to describe an object is known as a _____.
3. List the six views normally seen when making the drawing listed above.
    a. _____
    b. _____
    c. _____
    d. _____
    e. _____
    f. _____
4. The method used to develop these six views is called _____.

5. What views of the six views are normally used to describe an object?

OUTSIDE ACTIVITIES

1. Collect props for the class to draw, using instruments. One prop should require only a two-view drawing; another prop should require a three-view drawing. Find other props which need more than three views to give a complete shape description.
2. Build a hinged box out of clear plastic which can be used to demonstrate the unfolding of an object into its multiview parts; the front, top, bottom and sides. Place a prop inside the box, trace the profile of the object on the side of the plastic box with chalk, then unfold the box.
3. Make a large poster showing the step-by-step procedure to follow in centering a drawing on a sheet.

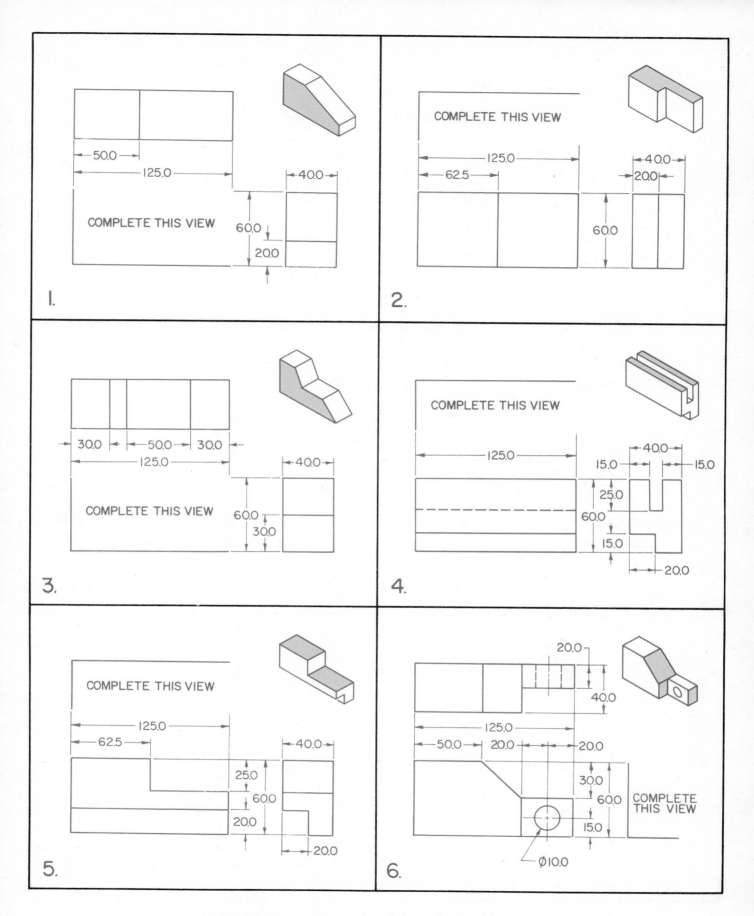

1.

50.0
125.0
40.0

COMPLETE THIS VIEW

60.0
20.0

2.

COMPLETE THIS VIEW

125.0
62.5
40.0
20.0

60.0

3.

30.0   50.0   30.0
125.0
40.0

COMPLETE THIS VIEW

60.0
30.0

4.

COMPLETE THIS VIEW

125.0
40.0
15.0   15.0
25.0
60.0
15.0
20.0

5.

COMPLETE THIS VIEW

125.0
62.5
40.0
25.0
60.0
20.0
20.0

6.

20.0
40.0
125.0
50.0   20.0   20.0
30.0
60.0   COMPLETE THIS VIEW
15.0
Ø10.0

PROBLEM SHEET 7–1. MULTIVIEW DRAWINGS. Draw each
problem on a separate sheet and complete as indicated.

99

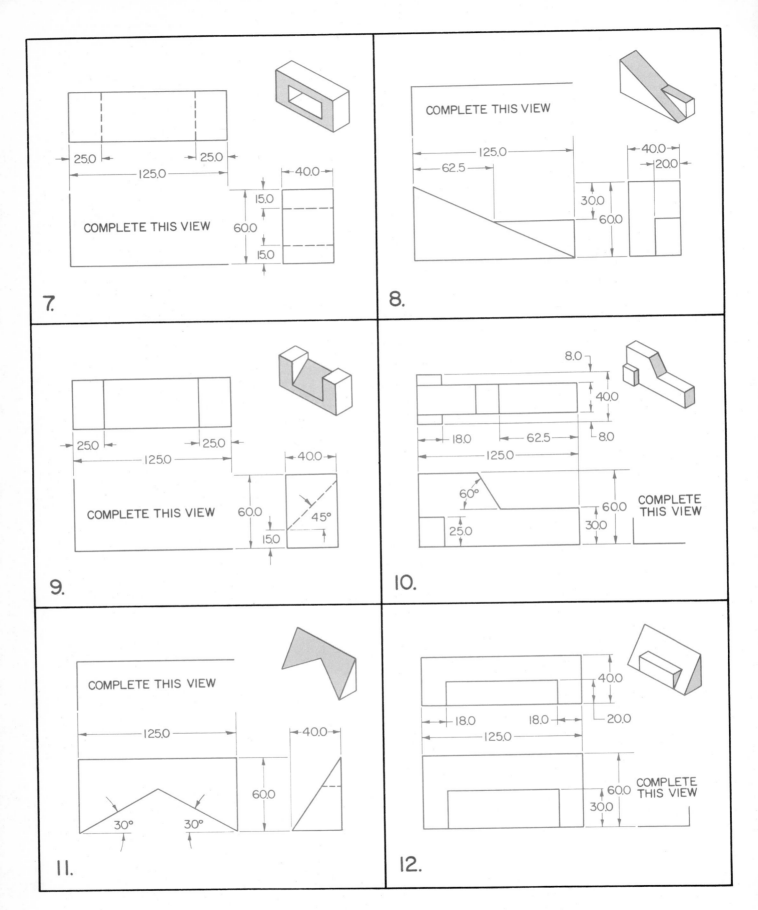

7.

25.0   25.0
125.0
40.0
15.0
60.0
15.0
COMPLETE THIS VIEW

8.

COMPLETE THIS VIEW
125.0
62.5
40.0
20.0
30.0
60.0

9.

25.0   25.0
125.0
40.0
60.0
45°
15.0
COMPLETE THIS VIEW

10.

8.0
40.0
18.0   62.5   8.0
125.0
60°
60.0
25.0   30.0
COMPLETE THIS VIEW

11.

COMPLETE THIS VIEW
125.0
40.0
60.0
30°   30°

12.

40.0
18.0   18.0   20.0
125.0
60.0
30.0
COMPLETE THIS VIEW

PROBLEM SHEET 7–2. MULTIVIEW DRAWINGS. Draw each
problem on a separate sheet and complete as indicated.

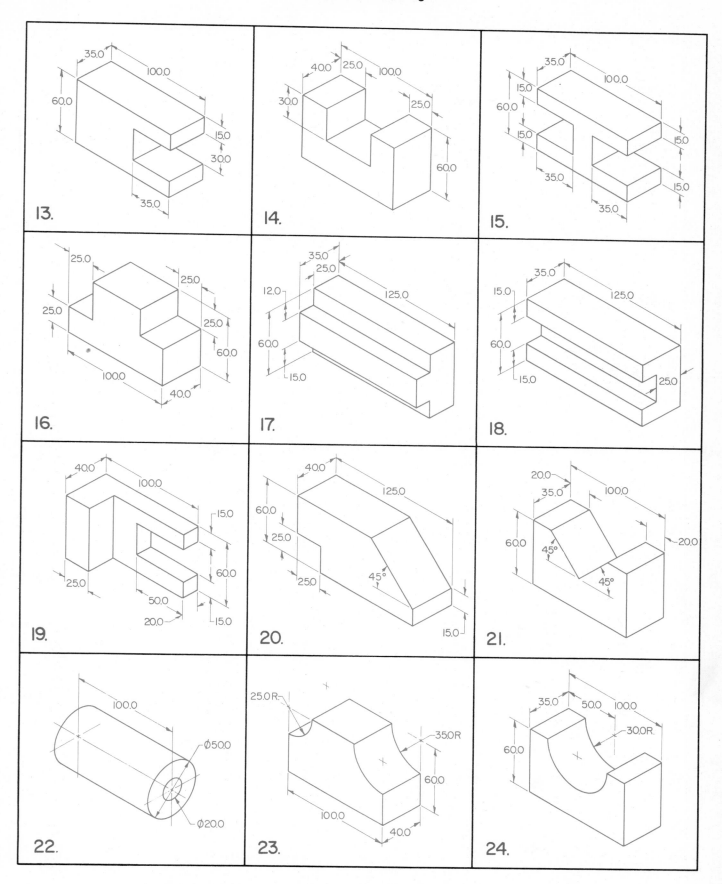

PROBLEM SHEET 7–3. MULTIVIEW DRAWINGS. Draw each problem on a separate
sheet. Draw as many views as necessary to fully describe each problem.

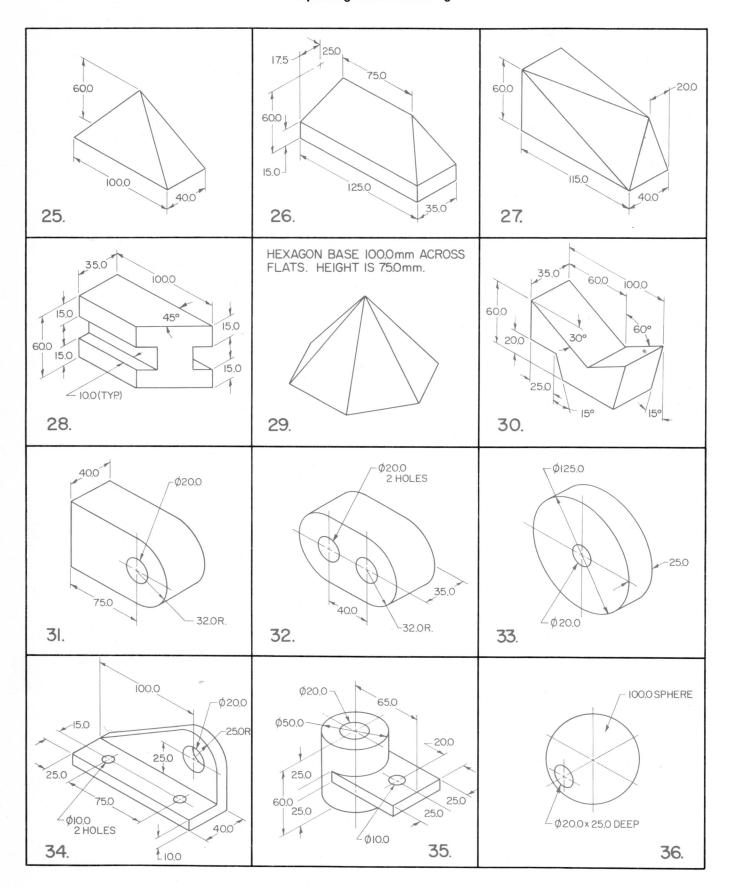

25. 60.0 100.0 40.0

26. 17.5 25.0 75.0 60.0 15.0 125.0 35.0

27. 60.0 20.0 115.0 40.0

28. 35.0 100.0 15.0 45° 15.0 60.0 15.0 15.0 10.0 (TYP.)

29. HEXAGON BASE 100.0 mm ACROSS FLATS. HEIGHT IS 75.0 mm.

30. 35.0 60.0 100.0 60.0 20.0 30° 60° 25.0 15° 15°

31. 40.0 Ø20.0 75.0 32.0 R.

32. Ø20.0 2 HOLES 35.0 40.0 32.0 R.

33. Ø125.0 25.0 Ø20.0

34. 100.0 Ø20.0 15.0 25.0 R 25.0 25.0 75.0 Ø10.0 2 HOLES 40.0 10.0

35. Ø20.0 65.0 Ø50.0 20.0 25.0 60.0 25.0 25.0 25.0 Ø10.0

36. 100.0 SPHERE Ø20.0 x 25.0 DEEP

PROBLEM SHEET 7—4. MULTIVIEW DRAWINGS. Draw each problem on a separate sheet. Draw as many views as necessary to fully describe each problem.

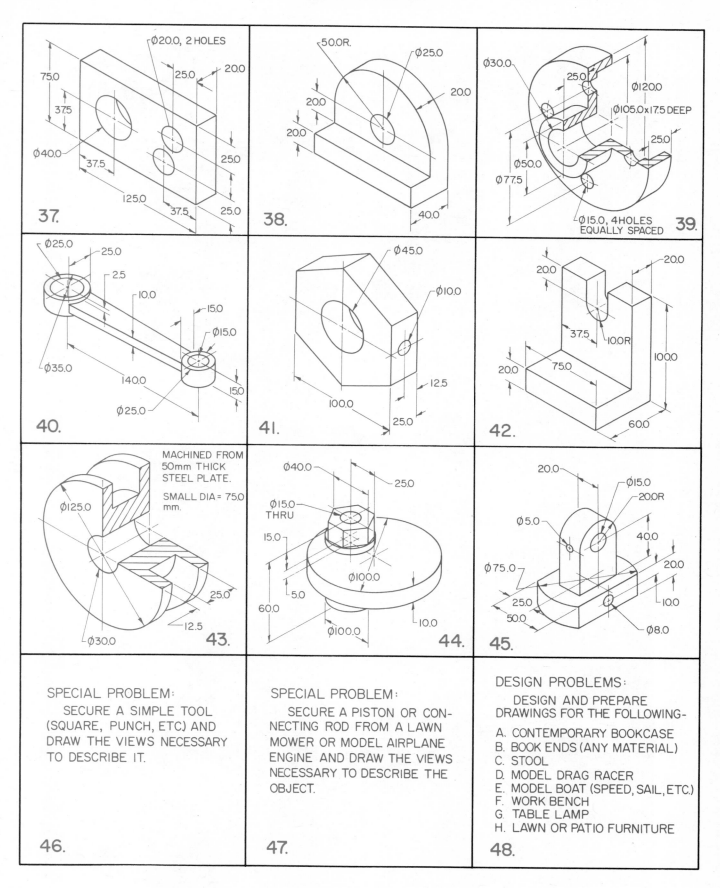

37. Ø20.0, 2 HOLES — 75.0 — 37.5 — Ø40.0 — 25.0 — 20.0 — 37.5 — 125.0 — 37.5 — 25.0 — 25.0

38. 50.0R. — Ø25.0 — 20.0 — 20.0 — 20.0 — 40.0

39. Ø30.0 — 25.0 — Ø120.0 — Ø105.0 x 17.5 DEEP — 25.0 — Ø50.0 — Ø77.5 — Ø15.0, 4 HOLES EQUALLY SPACED

40. Ø25.0 — 25.0 — 2.5 — 10.0 — 15.0 — Ø15.0 — Ø35.0 — 140.0 — 15.0 — Ø25.0

41. Ø45.0 — Ø10.0 — 12.5 — 100.0 — 25.0

42. 20.0 — 20.0 — 37.5 — 10.0R — 100.0 — 20.0 — 75.0 — 60.0

43. MACHINED FROM 50mm THICK STEEL PLATE. SMALL DIA = 75.0 mm. — Ø125.0 — 25.0 — 12.5 — Ø30.0

44. Ø40.0 — 25.0 — Ø15.0 THRU — 15.0 — 5.0 — Ø100.0 — 60.0 — Ø100.0 — 10.0

45. 20.0 — Ø15.0 — 20.0R — Ø5.0 — 40.0 — Ø75.0 — 20.0 — 25.0 — 10.0 — 50.0 — Ø8.0

46. SPECIAL PROBLEM: SECURE A SIMPLE TOOL (SQUARE, PUNCH, ETC.) AND DRAW THE VIEWS NECESSARY TO DESCRIBE IT.

47. SPECIAL PROBLEM: SECURE A PISTON OR CONNECTING ROD FROM A LAWN MOWER OR MODEL AIRPLANE ENGINE AND DRAW THE VIEWS NECESSARY TO DESCRIBE THE OBJECT.

48. DESIGN PROBLEMS: DESIGN AND PREPARE DRAWINGS FOR THE FOLLOWING—
A. CONTEMPORARY BOOKCASE
B. BOOK ENDS (ANY MATERIAL)
C. STOOL
D. MODEL DRAG RACER
E. MODEL BOAT (SPEED, SAIL, ETC.)
F. WORK BENCH
G. TABLE LAMP
H. LAWN OR PATIO FURNITURE

PROBLEM SHEET 7—5. MULTIVIEW DRAWINGS. Draw each problem on a separate sheet. Draw as many views as necessary to fully describe each problem.

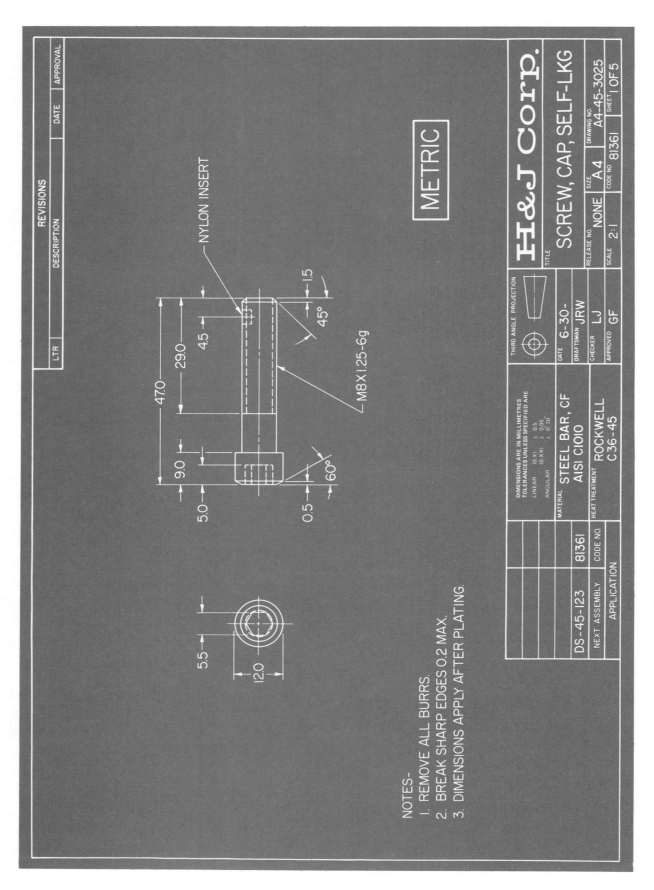

Fig. 8-1. Industry drawing showing dimensions and notes.

Unit 8
DIMENSIONING
AND SHOP NOTES

Many details are needed to manufacture an object according to the designer's specifications. A craftworker, for example, needs more information than that furnished by a scale drawing of its shape. DIMENSIONS and SHOP NOTES are needed. See Fig. 8-1.

DIMENSIONS define the sizes of the geometrical features of an object. They are presented in appropriate units of measure, such as millimetres, inches, etc.

SHOP NOTES provide additional information, along with sizes.

READING DIRECTIONS FOR DIMENSIONS

In the past, either UNIDIRECTIONAL or ALIGNED dimensioning was acceptable. See Fig. 8-2. Today, UNIDIRECTIONAL dimensioning is preferred and is used almost universally by industry.

Unidirectional dimensions are placed to be read from the bottom of the drawing. Leaders and notes are also lettered to read in the same way. In aligned dimensioning the numerals are placed parallel to their dimension lines. Aligned dimensions are read from the bottom and right side of the drawing.

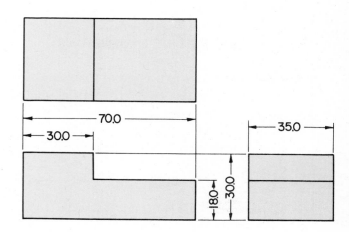

Fig. 8-2. Two accepted methods of dimensioning drawings. The UNIDIRECTIONAL method at left is preferred.

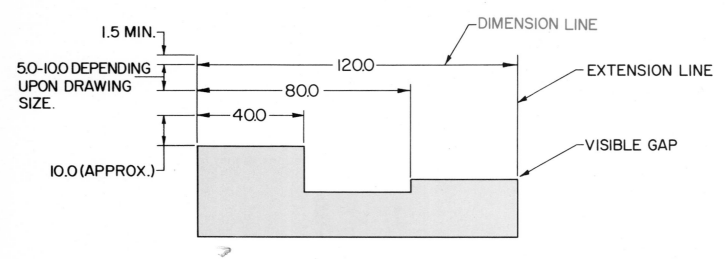

Fig. 8-3. Lines used for dimensioning. Note how they contrast with the object line.

DIMENSIONING A DRAWING

From your study of the ALPHABET OF LINES, you will remember that special lines are used for dimensioning, Fig. 8-3. The dimension line is a fine solid line used to indicate distance and location. It should be fine enough to contrast with the object lines. The line is broken near the center to receive the dimension. Arrowheads cap dimension line, Fig. 8-3.

The line that extends out from the drawing is called an extension line. It projects a minimum of 1.5 mm beyond the last dimension line. An extension line should not touch the drawing, Fig. 8-3. The smaller or detail dimensions are nearest the view. Larger or overall dimensions are farthest from the view.

Arrowheads, Fig. 8-4, are drawn freehand and should be carefully made. Since many industrial drawings are now stored on microfilm, the solid arrowhead is preferred. Arrowheads 3.0 mm long are satisfactory for most drawings.

HOW TO EXPRESS METRIC DIMENSIONS

Metric dimensions are expressed as follows:
1. The millimetre is the standard metric unit for dimensioning engineering drawings.

2. Millimetre dimensioning on engineering drawings is based on the use of one-place decimals; that is, one digit to the right of the decimal point. Two and three digits are used where critical tolerances are required. A zero will be shown to the right of the decimal point when full millimetre dimensions are shown.

125.0, not 125

3. When a dimension is less than one millimetre, a zero will be shown to the left of the decimal point.

0.5, not .5

4. All metric drawings should be clearly identified as such. The symbol for millimetre (mm) does not have to be added to every dimension if one of the notes shown in Fig. 8-5 is used on the drawing.

5. When the millimetre symbol is used on a drawing, a space should be placed between the dimension and the symbol.

125.0 mm, not 125.0mm

GENERAL RULES FOR DIMENSIONING

Dimensions must be easy to understand. They should conform to the following general rules:
1. Place dimensions on the views that show the true shape of the object, Fig. 8-6.
2. Unless absolutely necessary, dimensions should not be placed within the views.

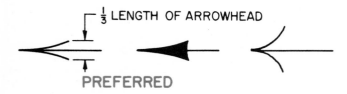

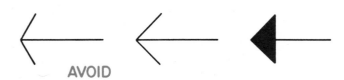

Fig. 8-4. Arrowheads are drawn freehand. The solid arrowhead generally is preferred.

METRIC

ALL DIMENSIONS ARE IN MILLIMETRES.

Fig. 8-5. Metric drawings should be clearly identified by use of either of the above notes.

3. If possible, dimensions should be grouped together. They should not be scattered about the drawing, Fig. 8-7.

4. Dimensions must be complete so that no scaling of the drawing is required. Also, it should be possible that sizes and shapes can be determined without assuming any measurements.

5. Draw dimension lines parallel to the direction of measurement. If there are several parallel dimension lines, stagger the numerals to make them easier to read. See Fig. 8-8.

6. Dimensions should not be duplicated unless they are absolutely necessary to the understanding of the drawing. Omit unnecessary dimensions, Fig. 8-9.

7. Plan your work carefully, so that dimension lines do not cross extension lines. See Fig. 8-10. Place the shortest lines next to the object outline.

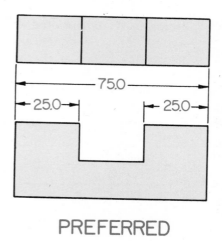

PREFERRED

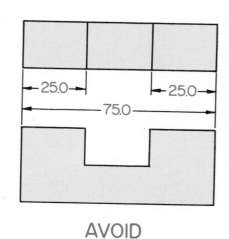

AVOID

Fig. 8-6. Dimension the view that shows the true shape of the object.

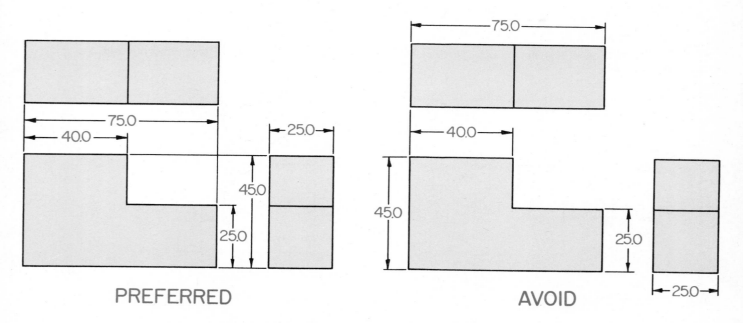

PREFERRED

AVOID

Fig. 8-7. Keep dimensions grouped for easier understanding of the drawing.

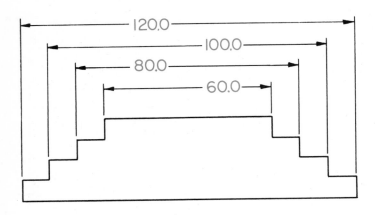

Fig. 8-8. When there are several parallel dimension lines, stagger the numerals.

DIMENSIONING CIRCLES, HOLES AND ARCS

The symbol Ø is used to indicate diameter. It precedes the dimension. The symbol DIA is preferred for use in narrative type notes.

Circles and round holes are dimensioned as shown in Fig. 8-11. The size represents the diameter. If the diameters of several concentric circles must be dimensioned on a drawing, it may be more convenient to show them on the front view, Fig. 8-12.

The correct way to use a leader to indicate a diameter is shown in Fig. 8-13. The leader is ALWAYS radial. That is, it points to the center of the diameter.

A circular arc is dimensioned by giving the radius, Fig. 8-14. The symbol "R" follows each dimension.

Round holes and cylindrical parts are dimensioned from centers, never from edges. See Fig. 8-15.

When it is necessary to dimension a series of holes around a circle, use a note to designate the number of holes, their size and the diameter of the circle. See notes in Fig. 8-16.

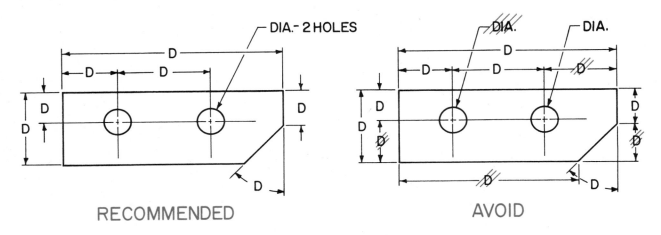

Fig. 8-9. Avoid duplicating dimensions unless they are necessary to the understanding of the drawing.

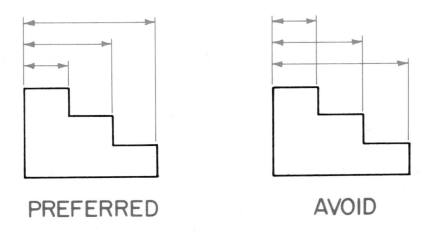

Fig. 8-10. Line crossing can be kept to a minimum by placing the shortest dimension lines next to the object outline.

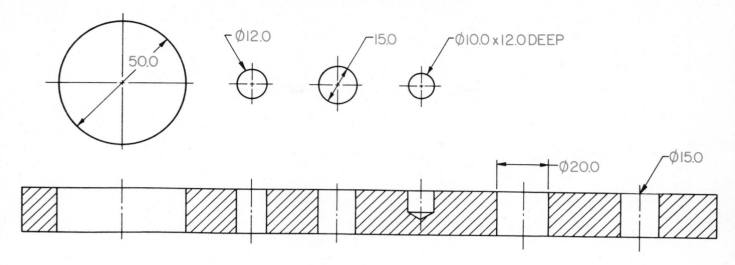

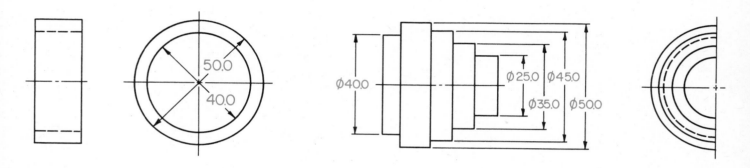

Fig. 8-11. Circles and holes are dimensioned by giving the diameter.

Fig. 8-12. Recommended ways to dimension concentric circles. Note that when the method at the right is used, only half of the right side view is needed.

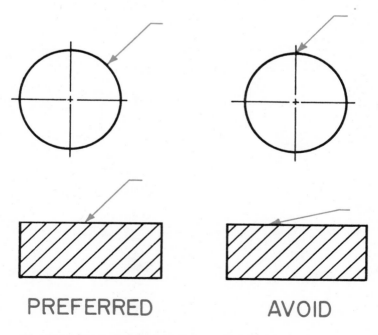

PREFERRED   AVOID

Fig. 8-13. Use a leader to direct attention to a note or to indicate sizes of arcs and circles. When used with arcs and circles, the leader should radiate from their centers.

Fig. 8-14. Recommended ways of dimensioning arcs.

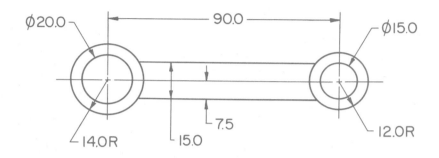

Fig. 8-15. Where feasible, dimension round holes and cylindrical parts from centers.

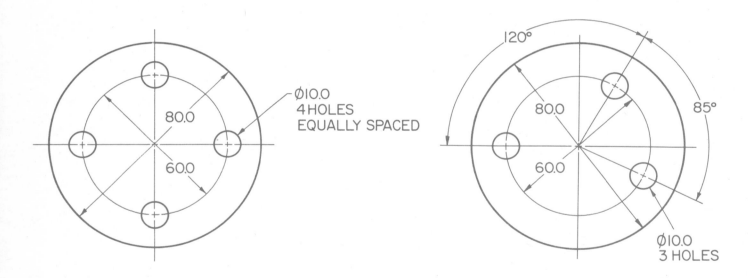

Fig. 8-16. Left. The correct way to dimension equally spaced holes. Right. Holes that are not equally spaced are dimensioned this way.

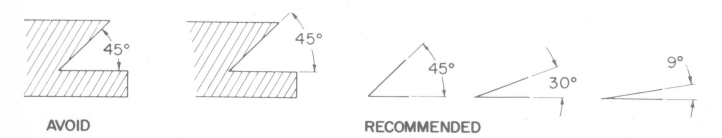

Fig. 8-17. Dimensioning angles.

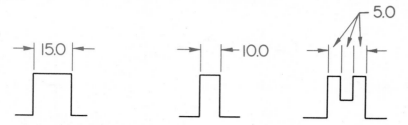

Fig. 8-18. When the space between extension lines is too small to place both arrowheads and numerals, indicate dimensions as shown.

DIMENSIONING ANGLES

Angular dimensions are expressed in degrees, minutes and seconds. Angles are dimensioned as shown in Fig. 8-17.

DIMENSIONING SMALL PORTIONS OF AN OBJECT

When the space between extension lines is too small to place both the numerals and the arrowheads, dimensions are indicated as shown in Fig. 8-18.

TEST YOUR KNOWLEDGE - UNIT 8

(Write answers on a separate sheet of paper.)
1. Dimensions and shop notes are needed to _____.
2. Dimensions define _____.
3. Shop notes provide _____.
4. Unidirectional dimensions are read from _____.
5. Aligned dimensions are read from _____.

6. The dimension line is a:
   a. Heavy solid line.
   b. Fine solid line.
   c. Fine dotted line.
   d. None of the above.
7. The dimension line is capped with _____.
8. Dimensions are placed on the views that show _____.
9. Dimensions should be _____ rather than _____ about the drawing.
10. Dimensions must be complete so that :
    a. No scaling of the drawing is necessary.
    b. Sizes and shapes can be determined without assuming any measurements.
    c. Both of the above.
    d. None of the above.
11. The symbol _____ is used to indicate diameter.
12. When a leader is used to indicate a diameter, it always points to the _____ of the diameter.

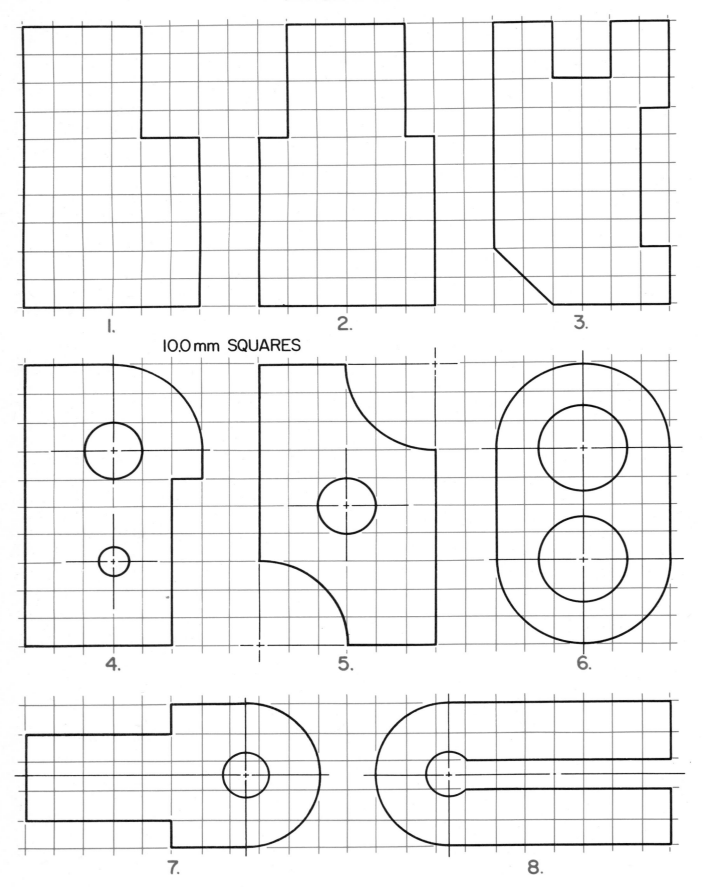

10.0 mm SQUARES

PROBLEMS 8—1 to 8—8. DIMENSIONING PROBLEMS. Redraw and dimension. Two of these problems should be drawn on a sheet.

10.0mm SQUARES

9.

10.

11.

12.

13.

14.

15.

PROBLEMS 8–9 to 8–15. DIMENSIONING PROBLEMS. Redraw and dimension problems.

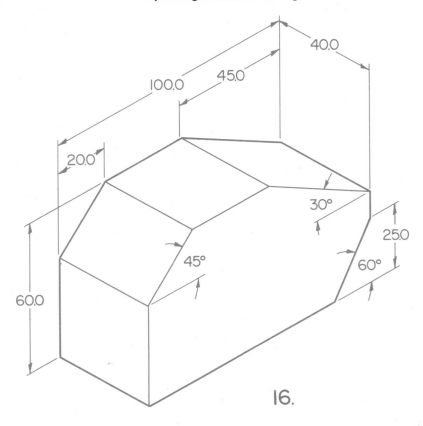

16.

PROBLEM 8–16. Above. SHIM. PROBLEM 8–17. Below. ALIGNMENT
PLATE. Prepare the necessary views and correctly dimension them.

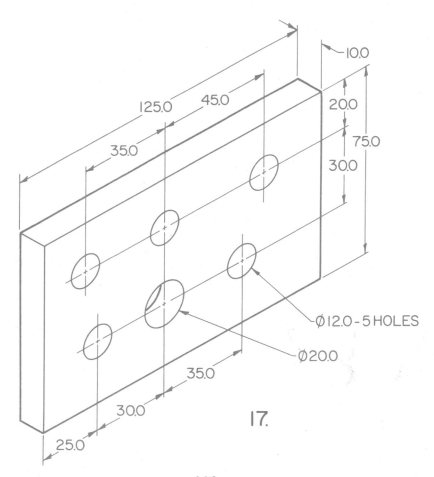

Ø12.0 - 5 HOLES

Ø20.0

17.

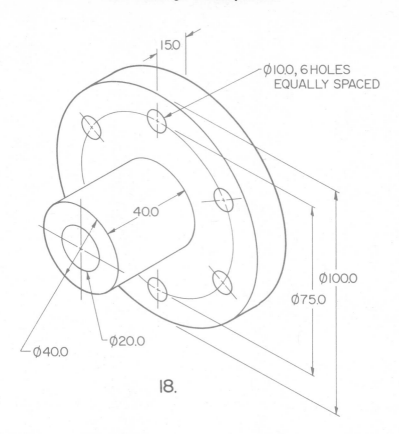

15.0

⌀10.0, 6 HOLES
EQUALLY SPACED

40.0

⌀100.0

⌀75.0

⌀20.0

⌀40.0

18.

PROBLEM 8–18. Above. FACE PLATE. Prepare a two view drawing and dimension correctly. PROBLEM 8–19. Below. COVER PLATE. Prepare the necessary views and dimension correctly.

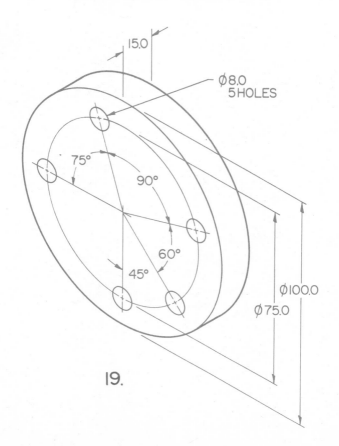

15.0

⌀8.0
5 HOLES

75°

90°

60°

45°

⌀100.0

⌀75.0

19.

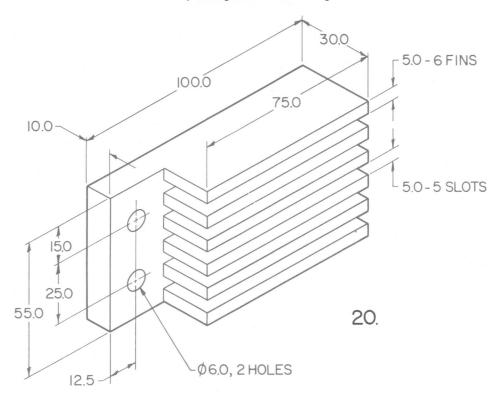

PROBLEM 8–20. Above. HEAT SINK. PROBLEM 8–21. Below. STEP
BLOCK. Prepare the necessary views and dimension correctly.

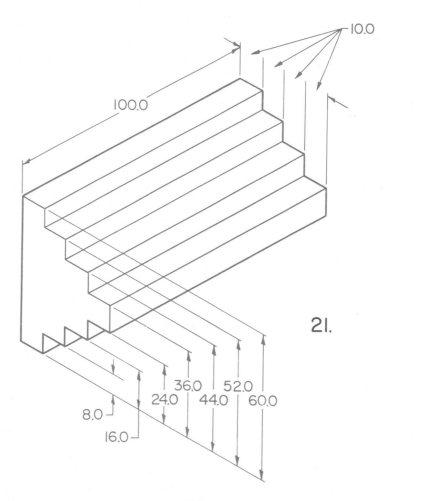

Unit 9
SECTIONAL VIEWS

Details on objects that are simple in design can be shown by using multiview projection. When an object has some of its design features hidden from view, as in Fig. 9-1, it is not easy to show the shape of the interior structure without using a "jumble" of hidden lines. SECTIONAL VIEWS are employed to make drawings of this type less confusing and easier to understand.

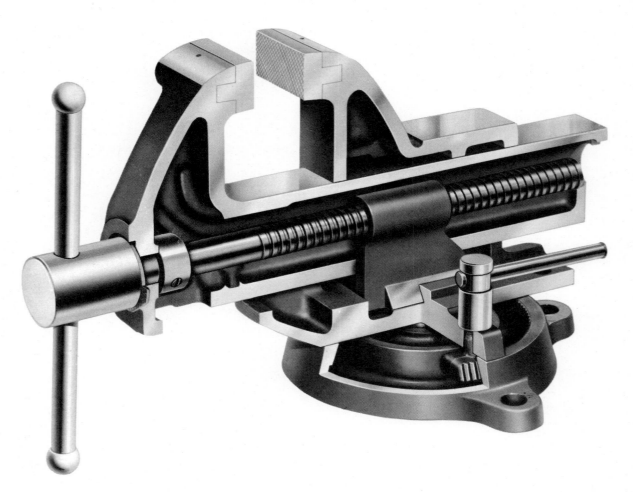

Fig. 9-1. Sectional view photo which shows the interior structure of a vise. (Columbian Vise and Mfg. Co.)

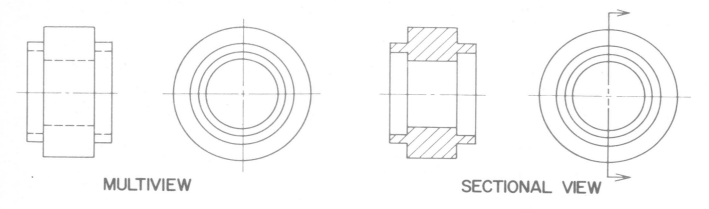

MULTIVIEW   SECTIONAL VIEW

Fig. 9-2. Conventional multiview drawing and a drawing in section.

SECTIONAL VIEWS show how an object would look if a cut were made through it perpendicular to the direction of sight, Fig. 9-2. Sectional view drawings are necessary for a clear understanding of the shape of complicated parts.

Fig. 9-3. Cutting plane lines. Either type may be used.

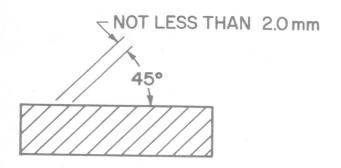

Fig. 9-4. General purpose section lining is spaced by eye and usually is drawn at 45 deg.

CUTTING PLANE LINE

The function of a CUTTING PLANE LINE, Fig. 9-3, is to indicate the location of an imaginary cut made through the object to reveal its interior characteristics. The extended lines are capped with arrowheads that indicate the direction of sight to view the section. Letters A—A, B—B, etc., are used to identify the section if it is moved to another position of the sheet, or if several sections are incorporated on a single drawing.

SECTION LINES

General purpose section lining (symbol for cast iron) is used where the material specifications ("specs") are lettered elsewhere on the drawing.

Spacing of general purpose section lining is by eye and usually at 45 deg., Fig. 9-4. Line spacing is somewhat dependent on the size of the drawing or the area to be sectioned. The lines usually are spaced about 2.0 mm apart for small areas, and up to 4.0 mm or more for large sections.

Do not use lines that are too thick or spaced too closely, Fig. 9-5. Also avoid lines that are not uniformly spaced or are drawn in different directions.

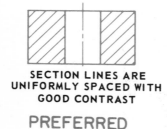

SECTION LINES ARE UNIFORMLY SPACED WITH GOOD CONTRAST

PREFERRED

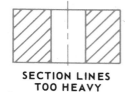

SECTION LINES TOO HEAVY

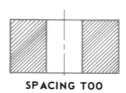

SPACING TOO CLOSE

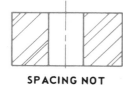

SPACING NOT UNIFORM

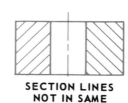

SECTION LINES NOT IN SAME DIRECTION

AVOID

Fig. 9-5. Section line spacing.

Fig. 9-6. The outline shape of the section may require the section lining to be drawn at other than 45 deg.

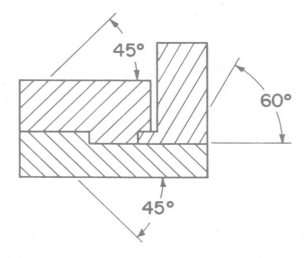

Fig. 9-7. Drawing section lines when several parts in the same sectional view are adjacent.

Where section lines will be parallel, or nearly parallel with the outline of the object, they should be drawn at some other angle, Fig. 9-6.

Should two or more pieces be shown in section, the section lines should be drawn in opposite directions or angles, to provide contrast, Fig. 9-7.

FULL SECTION

The FULL SECTION, Fig. 9-8, is developed by imagining that the cut has been made through the entire object. The part of the object between your eye and the cut is removed to reveal the interior features of the object. The resulting features are drawn as part of one of the regular multiview projections. Hidden lines behind the cutting plane are omitted unless they are necessary for a better understanding of the view or for dimensioning purposes.

HALF SECTION

The shape of one-half of the interior features and one-half of the exterior features of an object are shown in the HALF SECTION, Fig. 9-9. Half sections are best suited for symmetrical objects. Cutting plane lines are passed through the piece at right angles to each other. One-quarter of the object is considered removed to show a half section of the interior structure. Unless needed for clarity, hidden lines are eliminated from half sections.

REVOLVED SECTION

REVOLVED SECTIONS are primarily utilized to show the shape of such things as spokes, ribs and

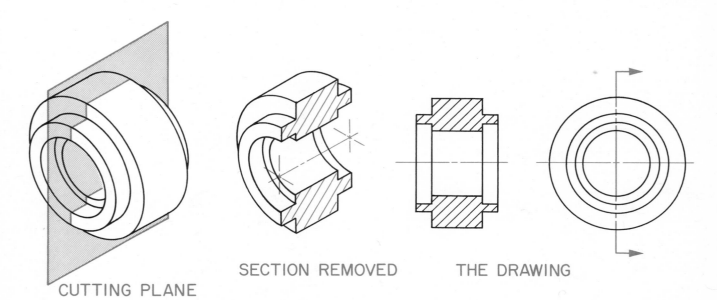

CUTTING PLANE SECTION REMOVED THE DRAWING

Fig. 9-8. The full section.

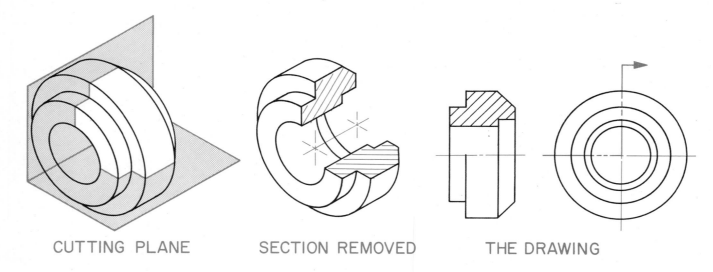

CUTTING PLANE   SECTION REMOVED   THE DRAWING

Fig. 9-9. The half section.

stock metal shapes, Fig. 9-10. A revolved section is a drawing within a drawing. To prepare a revolved section, imagine that a section of the part to be shown is cut out and revolved 90 deg. in the same view.

Do not draw the object lines of a normal view through the revolved section.

ALIGNED SECTION

It is not considered good practice to make a true full section of a symmetrical object that has an odd number of holes, webs or ribs. An ALIGNED SECTION is needed. With it, two of the divisions are

depicted on the view, one of them being revolved into the plane of the other, Fig. 9-11.

REMOVED SECTION

There are times when it is not possible to draw a needed sectional view on one of the regular views. When this situation occurs, a REMOVED SECTION is used, Fig. 9-12. The section (or sections) are placed elsewhere on the sheet.

Use the removed section when the section must be enlarged for better understanding of the drawing.

OFFSET SECTION

While the cutting plane is ordinarily taken straight through the object, it may be necessary to show features not located on the cutting plane. When this occurs, the cutting plane may be stepped or offset to

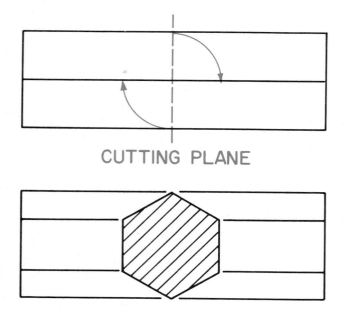

CUTTING PLANE

Fig. 9-10. A revolved section.

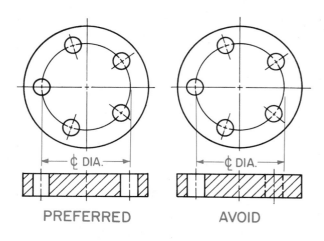

PREFERRED   AVOID

Fig. 9-11. An aligned section.

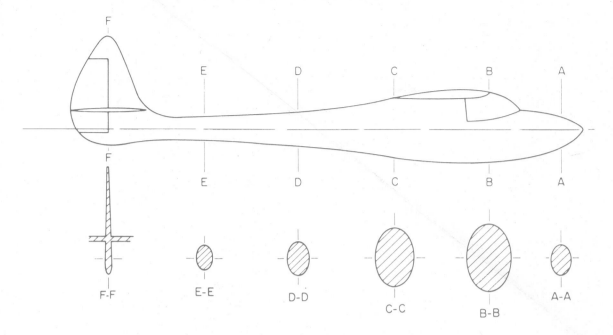

Fig. 9-12. A drawing showing the use of removed sections.

pass through these features, Fig. 9-13. This type of section is called an OFFSET SECTION.

The offsets are not shown in the sectional view. If necessary, use reference letters A—A, B—B, etc., at the ends of the cutting plane line.

BROKEN OUT SECTION

The BROKEN OUT SECTION is employed when a portion of a sectional view will provide required information. Break lines define the section and are shown on one of the regular views. See Fig. 9-14.

CONVENTIONAL BREAKS

When an object with a small cross section but of some length must be drawn, the drafter is faced with several problems. If drawn full size, it may be too long to fit on a standard size drawing sheet. If a scale small enough to fit the part on the sheet is used, the details may be too small to give the required information or to be dimensioned.

In such cases, the object can be fitted to the sheet by reducing its length by means of the CONVENTIONAL BREAK. See Fig. 9-15. This method per-

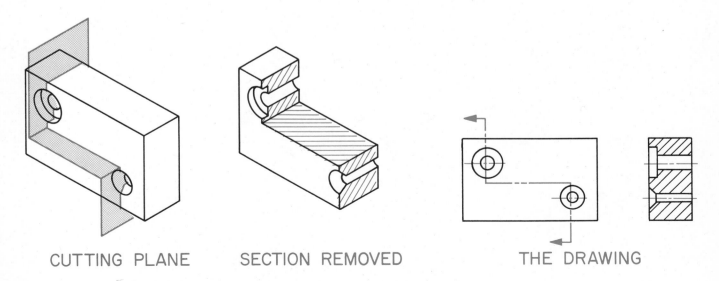

CUTTING PLANE   SECTION REMOVED   THE DRAWING

Fig. 9-13. An offset section.

121

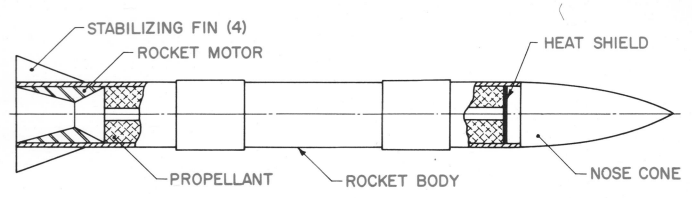

Fig. 9-14. Broken out section.

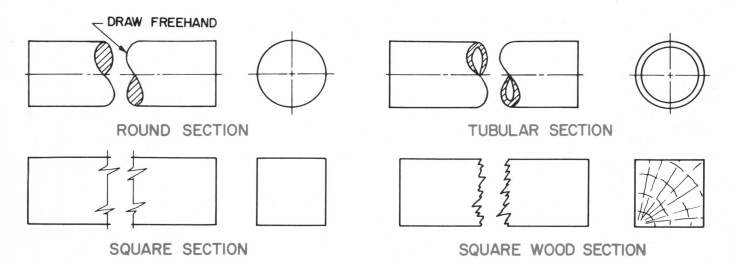

ROUND SECTION

TUBULAR SECTION

SQUARE SECTION

SQUARE WOOD SECTION

Fig. 9-15. Conventional breaks. By using breaks, an object with a small cross section but of some length may be drawn full size on a standard size drawing sheet.

mits a portion of the object to be deleted on the drawing.

Use the conventional break only when the cross section of the part is uniform its entire length. Note the different "break" techniques in Fig. 9-15.

SYMBOLS TO REPRESENT MATERIALS

Fig. 9-16 shows symbols which may be used on section views to indicate a number of different materials. The lines should be drawn dark and thin to contrast with the heavier object lines.

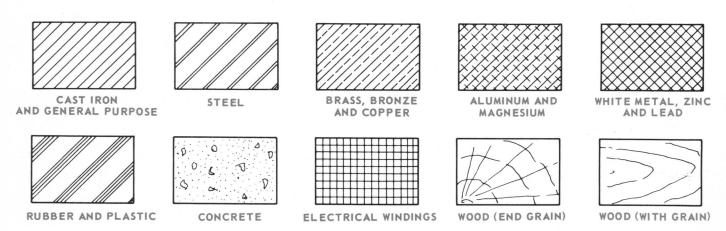

| CAST IRON AND GENERAL PURPOSE | STEEL | BRASS, BRONZE AND COPPER | ALUMINUM AND MAGNESIUM | WHITE METAL, ZINC AND LEAD |
| RUBBER AND PLASTIC | CONCRETE | ELECTRICAL WINDINGS | WOOD (END GRAIN) | WOOD (WITH GRAIN) |

Fig. 9-16. Standard code symbols to use for various materials shown in section.

TEST YOUR KNOWLEDGE - UNIT 9

(Write answers on a separate sheet of paper.)

1. What are sectional views?
2. When are sectional views used?
3. The _____ indicates location of imaginary cut made through object to show its interior details.
4. Sections are identified by the use of _____.
5. The _____ section is used when the cut has been made through the entire object.
6. The _____ section shows half of the interior and half of the exterior of the object.
7. _____ sections are primarily used to show the shape of such things as spokes, ribs and stock metal shapes.
8. The broken out section is used when _____.
9. Objects with small cross sections but of considerable length can be fitted on a standard size sheet by making use of the _____.
10. _____ may be used to indicate the material that has been cut.

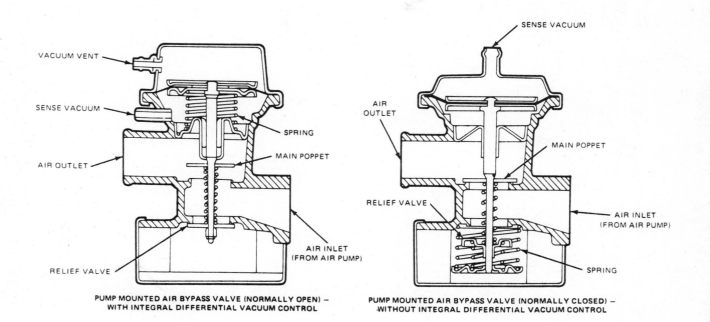

PUMP MOUNTED AIR BYPASS VALVE (NORMALLY OPEN) –
WITH INTEGRAL DIFFERENTIAL VACUUM CONTROL

PUMP MOUNTED AIR BYPASS VALVE (NORMALLY CLOSED) –
WITHOUT INTEGRAL DIFFERENTIAL VACUUM CONTROL

Cross-sectional views show differences in the construction of two types of air pump mounted by-pass valves used in Ford Motor Company's Thermactor air injection system. Note use of standard code symbol for cast iron.

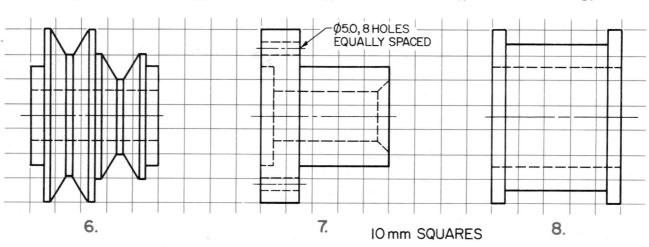

1.   2.   3.   4.   5.

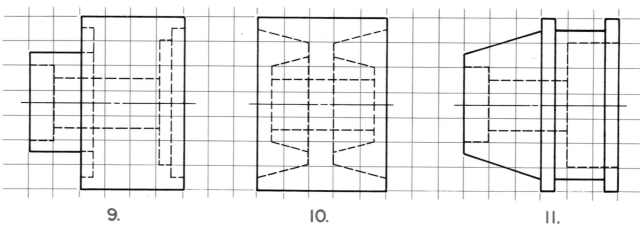

Ø5.0, 8 HOLES
EQUALLY SPACED

6.   7.   **10 mm SQUARES**   8.

9.   10.   11.

PROBLEMS 9–1 to 9–5. GRINDING WHEELS. Draw the views necessary to show the shape of each wheel. Make one view a full section or a half section as directed by your instructor. PROBLEM 9–6. PULLEY. Draw the views necessary to show the shape of this pulley. Draw one view as a half section. PROBLEM 9–7. COUPLING. Draw the views necessary to show the shape of this coupling. Draw one view as a full section. PROBLEM 9–8. BUSHING. Draw the views necessary to show the shape of this bushing. Draw one view as a half section. PROBLEM 9–9. SPACER. Draw the views necessary to show the shape of this spacer. Draw one view as a half section. PROBLEM 9–10. FLAT BELT PULLEY. Draw the views necessary to show the shape of this pulley. Draw one view as a full section. PROBLEM 9–11. ADAPTER BEARING. Draw the views necessary to show the shape of this bearing. Draw one view as a half section.

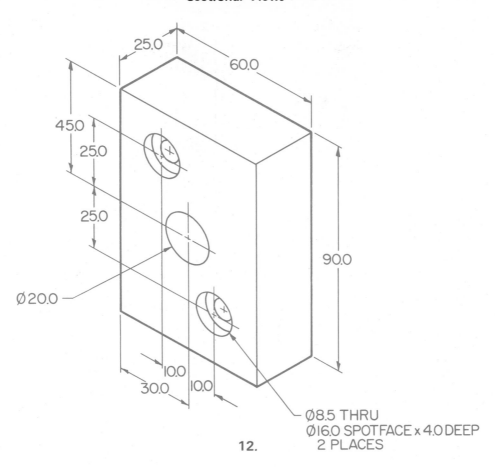

12.

PROBLEM 9–12. Above. SPACER. Draw the views necessary to show the shape of the spacer. Draw one view as an offset section through the three holes. PROBLEM 9–13. Below. SPECIAL BUSHING. Draw the views necessary to show the shape of the bushing. Draw one view as a full section along the horizontal plane.

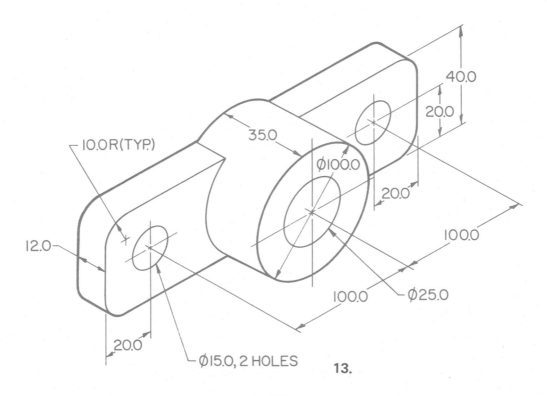

13.

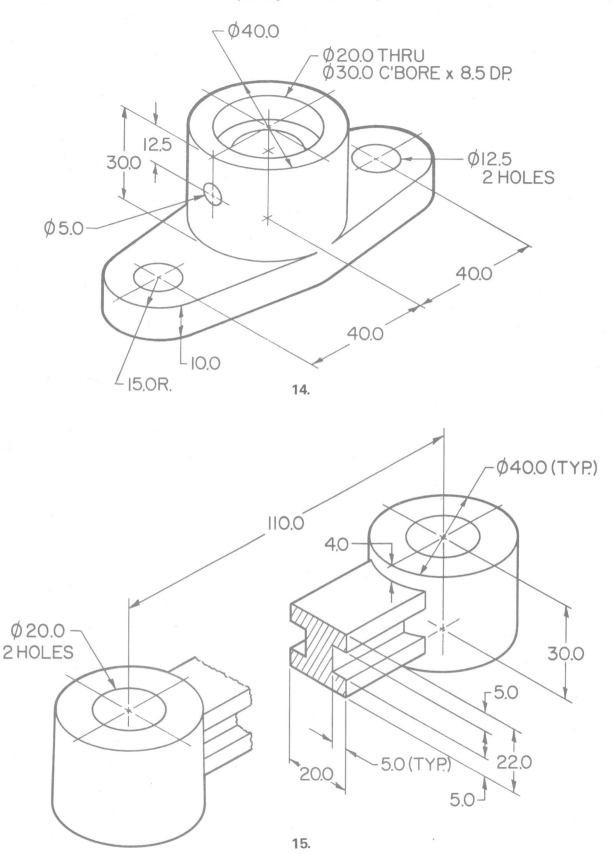

⌀40.0

⌀20.0 THRU
⌀30.0 C'BORE x 8.5 DP.

12.5

30.0

⌀12.5
2 HOLES

⌀5.0

40.0

40.0

10.0

15.0R.

14.

⌀40.0 (TYP.)

110.0

4.0

⌀20.0
2 HOLES

30.0

5.0

5.0 (TYP.)

20.0

22.0

5.0

15.

PROBLEM 9–14. BRACKET. Draw the views necessary to show the shape of the bracket. Include a broken out section through the 5.0 mm diameter hole on one view. PROBLEM 9–15. CONNECTING ROD. Draw the views necessary to show the shape of the rod. The cross section of the rod may be shown as a revolved section or as a removed section.

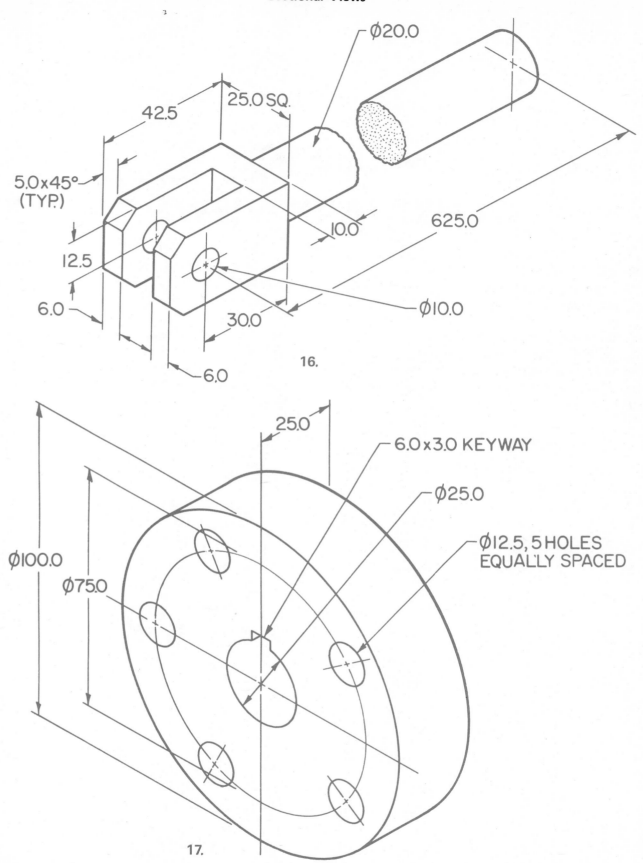

Ø20.0

42.5

25.0 SQ.

5.0x45°
(TYP.)

12.5

6.0

30.0

6.0

10.0

625.0

Ø10.0

16.

25.0

6.0x3.0 KEYWAY

Ø25.0

Ø12.5, 5 HOLES
EQUALLY SPACED

Ø100.0

Ø75.0

17.

PROBLEM 9–16. TORQUE ROD. Draw the views necessary to show the shape of the rod. Use the conventional break to fit the object on the drawing sheet. PROBLEM 9–17. ADAPTER PLATE. Draw the views necessary to show the shape of the plate. Draw one view as a full section.

Unit 10
AUXILIARY VIEWS

The true shape and size of objects having angular or slanted surfaces cannot be shown using the regular (top, front, side) views. The TRUNCATED BLOCK shown in Fig. 10-1 is such an object (truncated means that the object has been cut off at an angle). The true length of the cut is shown on the front view, but this view does not show its width. The true width of the cut is shown on the top and side views but neither shows the true length.

An additional or AUXILIARY VIEW is needed to show the true length and true width of the angular surface, Fig. 10-2.

When drawing an auxiliary view, remember that the view is ALWAYS projected from the regular view on which the inclined surface appears as a line. Also, the construction lines projecting from the inclined surface are ALWAYS at right angles to the cut.

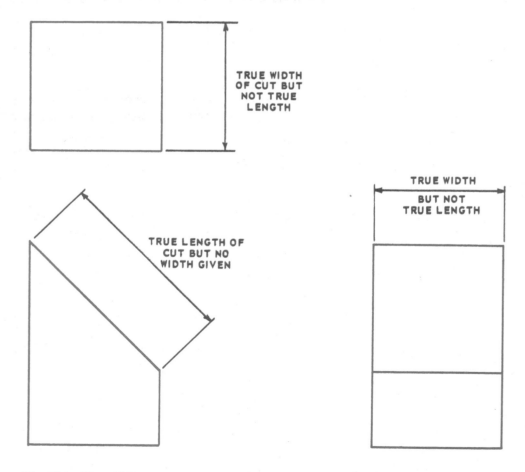

TRUE WIDTH OF CUT BUT NOT TRUE LENGTH

TRUE LENGTH OF CUT BUT NO WIDTH GIVEN

TRUE WIDTH BUT NOT TRUE LENGTH

Fig. 10-1. Why AUXILIARY VIEWS are necessary. The single views do not show the true shape (length and width) of the cut off portion of the object.

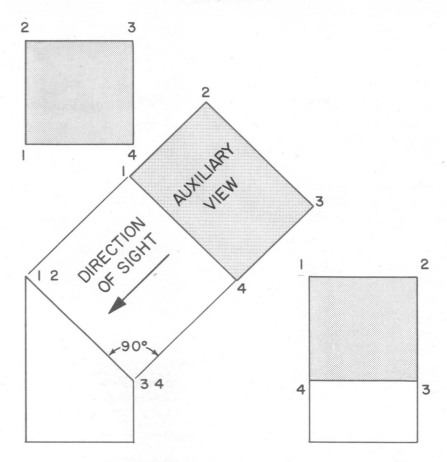

Fig. 10-2. The AUXILIARY VIEW shows the true shape of the angular surface.

When drawing auxiliaries, the usual practice is to show only the inclined portion of the view. It is seldom necessary to draw a full projection of the object, Fig. 10-3. It is often possible to eliminate one of the conventional views when using an auxiliary view. See Fig. 10-4.

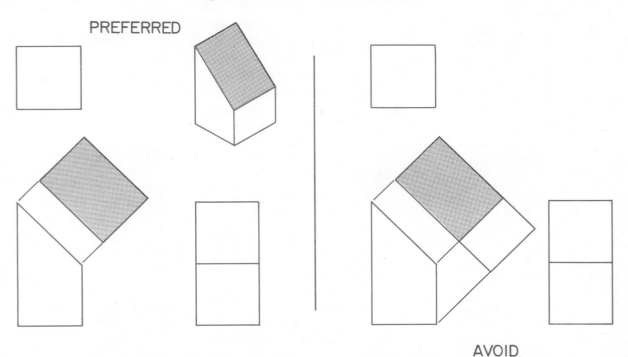

Fig. 10-3. It is not necessary to draw a full projection of the object.

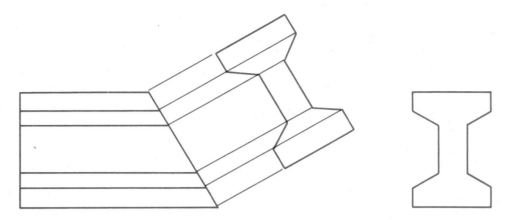

Fig. 10-4. One view may often be eliminated when using an auxiliary view.

An auxiliary view can be projected from any view that shows the inclined surface as a line. It would be called a FRONT AUXILIARY if projected from the front view, a TOP AUXILIARY if projected from the top view, and a RIGHT SIDE AUXILIARY if projected from the right side.

Considerable time can be saved when drawing auxiliary views of symmetrical objects by drawing only half of the view.

Auxiliary views that include rounded surfaces or circular openings may cause minor problems. Fig. 10-5 shows how such a rounded surface would be drawn. Proceed as follows:

1. Draw the needed views. Divide the circular view into 12 equal parts. Project the divisions to the other view.
2. At any convenient distance from the inclined face, draw center line A'—A' for the auxiliary view. This center line is parallel to the inclined face.
3. Project the necessary points from the inclined surface through center line A'—A'. These lines are at right angles (90°) to the inclined face.
4. Using dividers or a compass, transfer measurements A—a, A—b and A—c to the appropriate lines to the right and left of center line A'—A' (to get points A'—a, A'—b and A'—c.
5. Complete the auxiliary view by connecting the points with a French curve.

TEST YOUR KNOWLEDGE - UNIT 10

(Write answers on a separate sheet of paper.)
1. Why are auxiliary views needed?
2. Sketch an object that requires an auxiliary view.
3. The auxiliary view is always projected from the view that shows the inclined surface as a _____. The construction lines projecting from the inclined surface are always at _____ to the cut.
4. The auxiliary view when projected from the front view is called a _____.
5. When drawing an auxiliary view of a symmetrical object, much time can be saved by drawing _____.

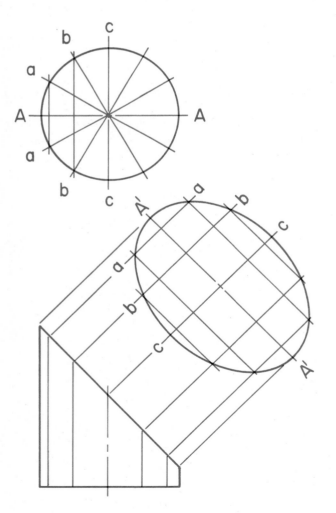

Fig. 10-5. Drawing an auxiliary view of a circular object.

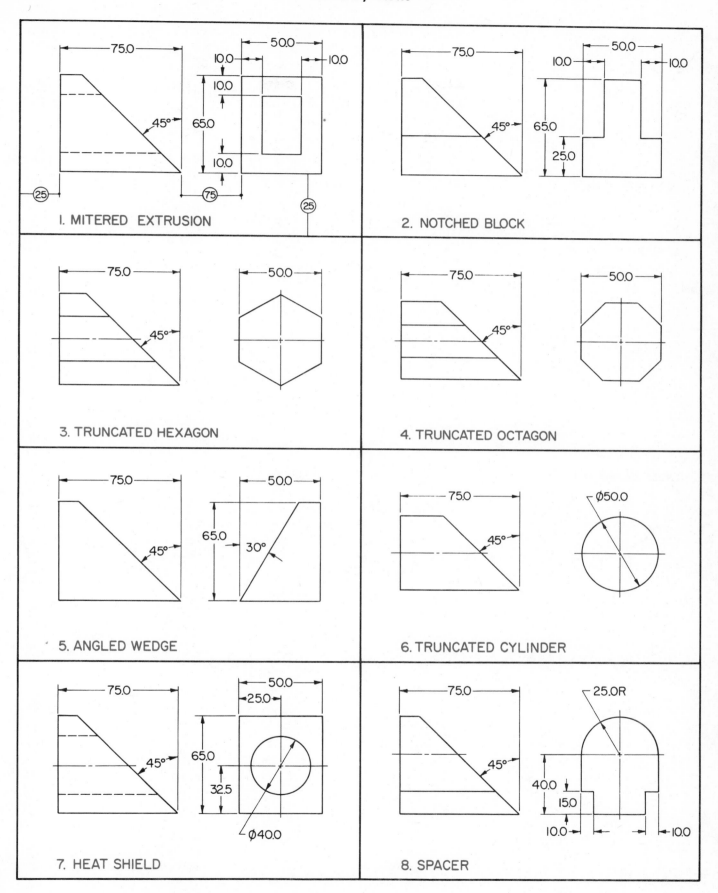

1. MITERED EXTRUSION

2. NOTCHED BLOCK

3. TRUNCATED HEXAGON

4. TRUNCATED OCTAGON

5. ANGLED WEDGE

6. TRUNCATED CYLINDER

7. HEAT SHIELD

8. SPACER

PROBLEMS 10–1 to 10–8. Space drawings according to the circled dimensions. The top views may be eliminated.

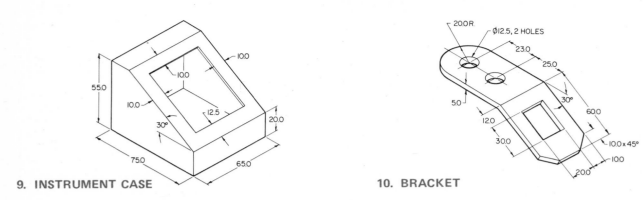

9. INSTRUMENT CASE

10. BRACKET

PROBLEM 10—9. Left. INSTRUMENT CASE. 1—Use a vertical drawing sheet format. 2—Allow 100 mm between front and top views. 3—Locate the auxiliary view 35 mm from the front view. 4—The front view is 25 mm from the left border. PROBLEM 10—10. Right. BRACKET. 1—Use a horizontal drawing sheet format. 2—Allow 75 mm between the front and top views. 3—Locate the auxiliary view 25 mm from the front view. 4—The front view is 70 mm from the left border.

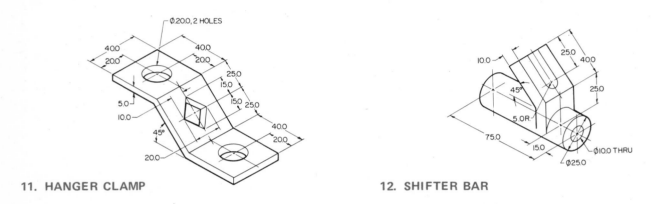

11. HANGER CLAMP

12. SHIFTER BAR

PROBLEM 10—11. Left. HANGER CLAMP. 1—Use a horizontal drawing sheet format. 2—Allow 75 mm between front and top views. 3—Locate the auxiliary view 35 mm from the front view. 4—Allow 60 mm between front and right side view. 5—The front view is 20 mm from the left border. PROBLEM 10—12. Right. SHIFTER BAR. 1—Use a horizontal drawing sheet format. 2—Allow 50 mm between the front and top views. 3—Locate the auxiliary view 25 mm from the front view. 4—Allow 50 mm between the front and right view. 5—The front view is 50 mm from the left border.

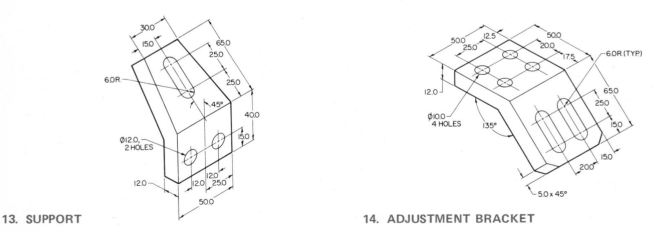

13. SUPPORT

14. ADJUSTMENT BRACKET

PROBLEM 10—13. Left. SUPPORT. 1—Use a vertical sheet format. 2—Allow 65 mm between front and top views. 3—Locate the auxiliary view 35 mm from the front view. 4—The front view is 30 mm from the left border. PROBLEM 10—14. Right. ADJUSTABLE BRACKET. 1—Use a horizontal sheet format. 2—Allow 55 mm between the front and top views. 3—Locate the auxiliary view 25 mm from the front view. 4—Allow 70 mm between the front and right side views. 5—The front view is 20 mm from the left border.

Ford Motor Company's four cylinder, 2.3 litre power plant is the first metric
engine designed, developed and manufactured in the U.S.

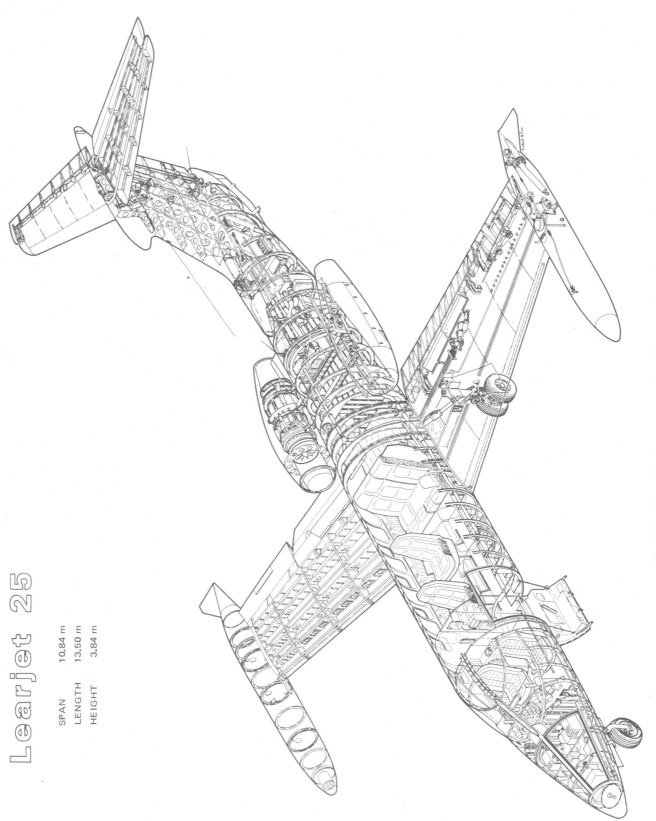

Learjet 25

SPAN	10.84 m
LENGTH	13.50 m
HEIGHT	3.84 m

Fig. 11-1. Pictorials have many uses. This one shows the structural details of a jet plane. (Gates—Learjet)

Unit 11
PICTORIALS

PICTORIAL DRAWINGS

A PICTORIAL DRAWING shows a likeness (shape) of an object as viewed by the eye. The pictorial drawing of the jet plane, Fig. 11-1, shows many of the structural details of the craft.

If you have worked with radio or electronic kits, you are familiar with pictorial drawings which show the builder what needs to be done and how to do it to complete his project, Fig. 11-2.

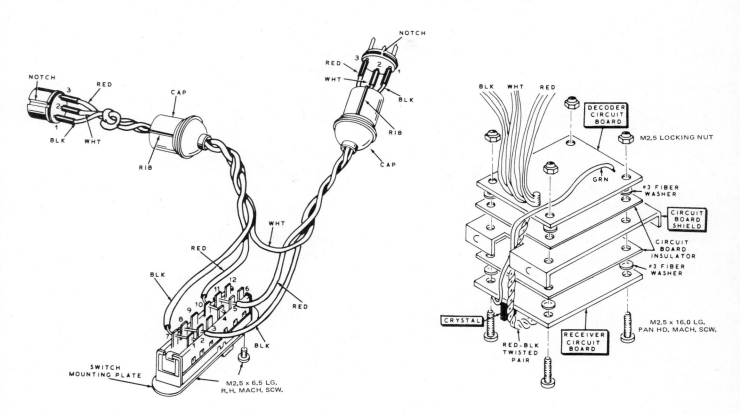

Fig. 11-2. Pictorials are used to give instructions in many hobby areas. These are from a radio construction manual. (Heath Co.)

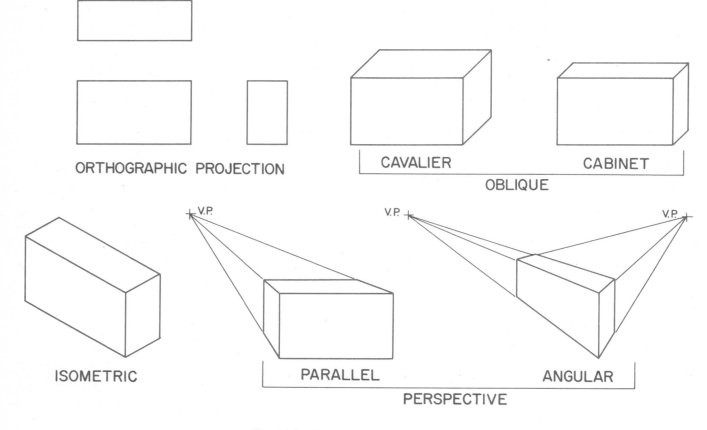

ORTHOGRAPHIC PROJECTION

CAVALIER   CABINET
OBLIQUE

ISOMETRIC   PARALLEL   ANGULAR
PERSPECTIVE

Fig. 11-3. Types of pictorial drawings.

Several types of pictorial drawings with which you should become familiar are shown in Fig. 11-3.

ISOMETRIC DRAWINGS

An ISOMETRIC DRAWING shows an object as it is. All lines which show the width and depth are drawn full length (or in the same proportionate length). Edges which are upright are shown by vertical lines. The object is assumed to be in position with its corners toward you, and its horizontal edges sloping away at angles of 30 deg. to the right and to left. NOTE: Hidden lines are omitted from isometric drawings unless needed for clarity.

An ISOMETRIC DRAWING is made entirely with instruments, using three base lines, Fig. 11-4. One of the lines is vertical, the other two are drawn at an angle of 30 deg. to the horizontal. The base lines may be reversed if more information can be shown with the object in the reversed position, Fig. 11-5.

Lines not parallel to the three base lines are called NON—ISOMETRIC LINES. These may be drawn by transferring reference points from a multiview drawing to the isometric drawing, Fig. 11-6.

Isometric circles, Fig. 11-7, (circles which are not true ellipses) may be drawn using a compass and a 60

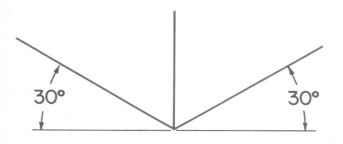

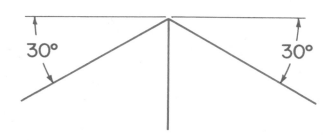

Fig. 11-4. Lines required to make isometric drawings.

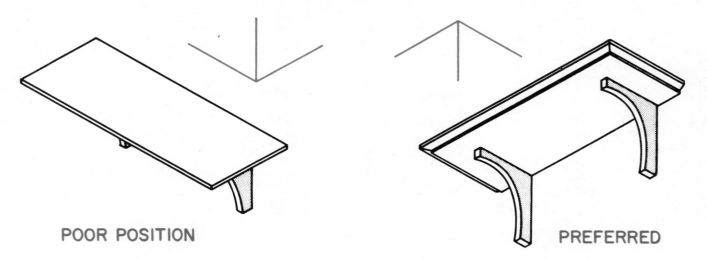

POOR POSITION

PREFERRED

Fig. 11-5. Reverse the base lines if more information can be shown.

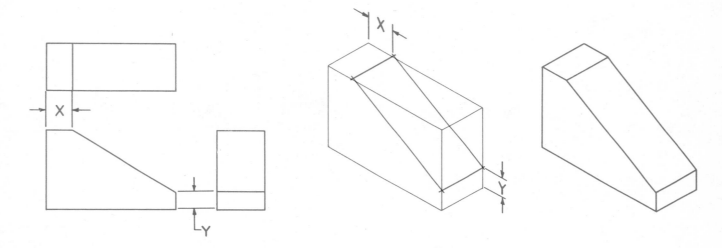

Fig. 11-6. Drawing non-isometric lines.

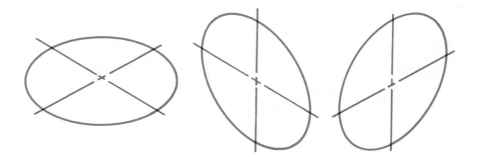

Fig. 11-7. Isometric circles.

deg. triangle, as shown in the two examples in Fig. 11-8. First, make an isometric square into which the isometric circle is to be drawn, Detail A. Using a 60 deg. triangle, draw lines as in Detail B. This provides four centers from which arcs may be drawn tangent (line which contacts other lines at only one point) to sides of the isometric square. The four centers are marked by small circles in Details C and D.

It is well to know how to draw isometric circles using drafting tools, but an easier way is to use an ellipse template of appropriate size. See Fig. 11-9.

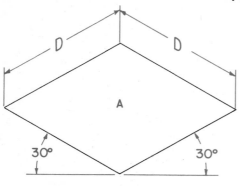

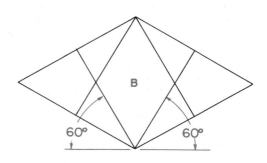

D= DIAMETER OF
REQUIRED CIRCLE

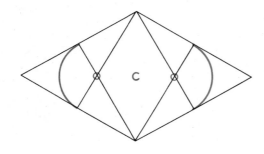

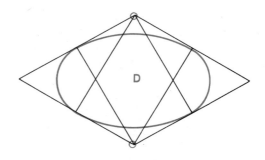

Fig. 11-8. Drawing isometric circles with a compass.
Above. On a flat plane. Below. On a vertical plane.

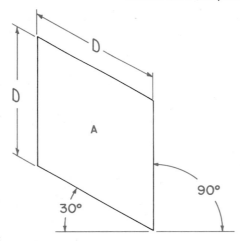

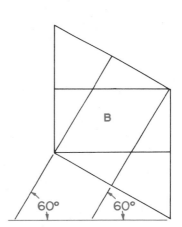

D= DIAMETER OF
REQUIRED CIRCLE

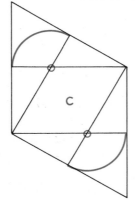

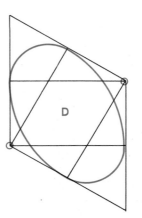

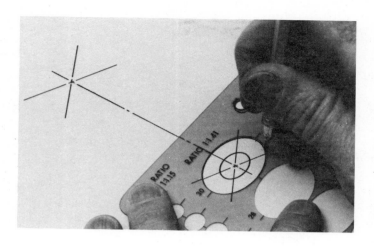

Fig. 11-9. Using an isometric circle template.

OBLIQUE PICTORIAL DRAWINGS

An OBLIQUE DRAWING is a pictorial drawing similar to an isometric drawing. However, one surface of an oblique drawing, the longest dimension or front, is parallel to the picture plane (is shown in true shape and size).

Three types of oblique drawings are shown in Fig. 11-10. Each type shows the front of the object as it appears in multiview form. The angle of the depth axis (construction lines drawn at an angle) may be any angle, but 15, 30 or 45 deg. are generally used.

CAVALIER OBLIQUE. An oblique drawing in which the depth axis lines are full scale (full size).

CABINET OBLIQUE. Depth axis lines are drawn one-half scale.

GENERAL OBLIQUE. Depth axis lines vary from one-half to full scale.

DIMENSIONING PICTORIAL DRAWINGS

To dimension isometric and oblique drawings, draw dimension lines parallel to the corresponding planes of the object (lines of the drawing). See Fig. 11-11. Acceptable alternate dimensioning is also shown in Fig. 11-11.

PERSPECTIVE DRAWINGS

A PERSPECTIVE DRAWING is a drawing used by an architect, artist or drafter to show an object as it would appear to the eye from a certain location.

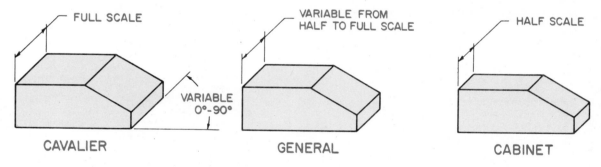

Fig. 11-10. Types of oblique drawings. The CAVALIER and CABINET drawings are most frequently drawn.

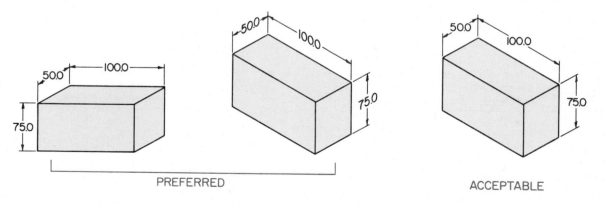

Fig. 11-11. Dimensioning pictorials.

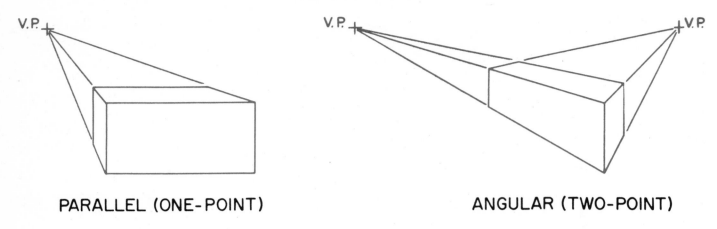

PARALLEL (ONE-POINT)   ANGULAR (TWO-POINT)

Fig. 11-12. Perspective drawings.

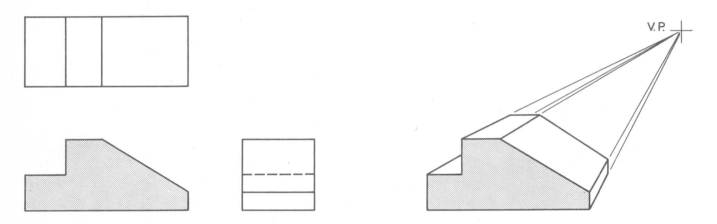

Fig. 11-13. On PARALLEL or ONE-POINT PERSPECTIVE DRAWINGS, the front view is shown in its true shape and in full or scale size.

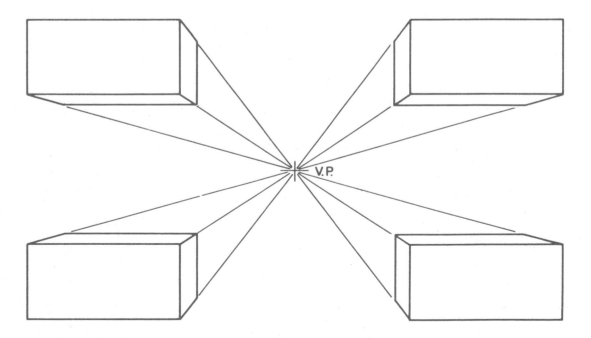

Fig. 11-14. The perspective may be located in any position relative to the vanishing point (V.P.).

To understand how a perspective is formed, assume that you are standing between two railroad tracks, in the exact center of the space, and that the tracks run straight toward the horizon as far as you can see. As you look down the tracks, the tracks appear to converge (run together) at an eye level point on the horizon. The point where the tracks meet is called the VANISHING POINT (V.P.). Lines above eye level seem to recede down to the horizon, while those below eye level appear to rise upward to the horizon. In making a perspective drawing, lines used to show depth or shape of the object (if continued) will converge to form a V.P., Fig. 11-12.

To make simplified ONE—POINT or PARALLEL PERSPECTIVE DRAWING, Fig. 11-13, draw the front view in its true shape in full or scale size. Draw the horizon line and select the VANISHING POINT (V.P.) at random. If at first you do not obtain a pleasing effect, try another vanishing point location. Information on locating vanishing points more accurately may be obtained from a text covering advanced drafting procedures.

The perspective may be located in any position relative to the V. P., Fig. 11-14. Project construction lines from each point on the front view to the vanishing point, Fig. 11-15.

Simple ANGULAR or TWO—POINT PERSPECTIVES may be drawn as shown in Fig. 11-16.

In drawing perspectives, considerable saving of time and additional accuracy will result if you use a perspective drawing board. See Fig. 11-17. This type board is available from concerns that sell school and drafting supplies.

EXPLODED ASSEMBLY DRAWINGS

EXPLODED ASSEMBLY DRAWINGS are easy to read and may be understood without having an extensive knowledge of blueprint reading. See Fig. 11-18.

Such drawings are used extensively by industry and in many hobby areas. You are probably quite familiar with this type of drawing if you build model cars, planes, boats or rockets.

Exploded assembly drawings are nothing more than a series of pictorial drawings (usually isometrics)

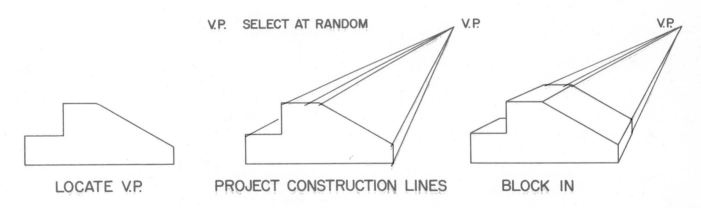

LOCATE V.P.   PROJECT CONSTRUCTION LINES   BLOCK IN

Fig. 11-15. Drawing a one-point perspective.

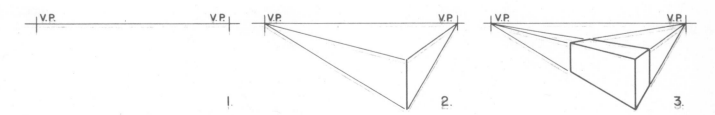

Fig. 11-16. Drawing a simple ANGULAR or TWO-POINT PERSPECTIVE: 1—Locate the position of the horizon and two vanishing points. 2—Locate and draw a vertical line the full or scale height of the object being drawn. 3—Draw construction lines from the vertical line to the vanishing points. For simple two-point perspectives like this, it is permissible to locate the length and depth in a position that gives the most pleasing effect. Darken the necessary lines.

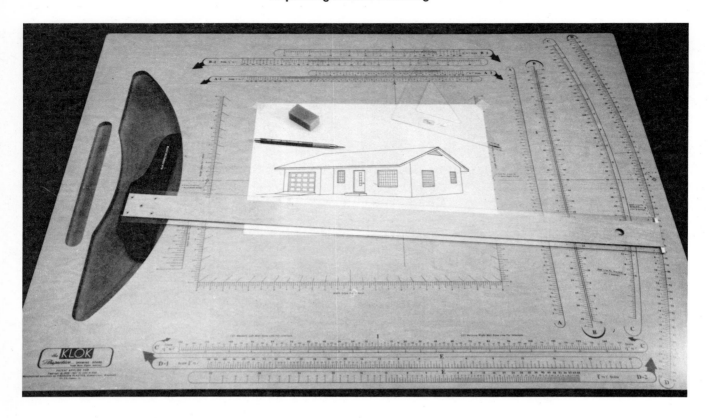

Fig. 11-17. Klok perspective drawing board.

showing the parts that make up the object. They are located in proper relation to each other.

HOW TO CENTER PICTORIAL DRAWINGS

There are several ways to center pictorial drawings on the drawing sheet.

One method widely used, is called "eye balling." You estimate by sight the approximate location of

the starting point and develop your drawing from this point.

Another method requires the drawing of the pictorial on a second piece of paper. When it has been checked for accuracy, measure its total width and height. With this information, locate the starting point of the drawing so it will be centered on the drawing sheet which is to be evaluated.

A third method, one that will work on most problems, is shown in Fig. 11-19.

TEST YOUR KNOWLEDGE - UNIT 11

(Write answers on a separate sheet of paper.)

1. Pictorial drawing is a method of showing an object as:
   a. It would look on the drawing.
   b. It would appear to the eye.
   c. Another form of multiview drawing.
   d. All of the above.
   e. None of the above.
2. Why are pictorials often used?
3. List 4 types of pictorial drawings:
   a. _____.
   b. _____.

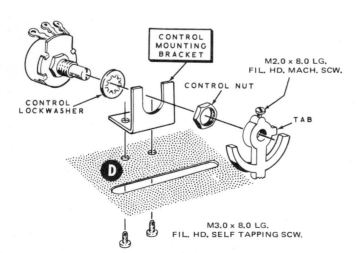

Fig. 11-18. An exploded pictorial drawing. (Heath Co.)

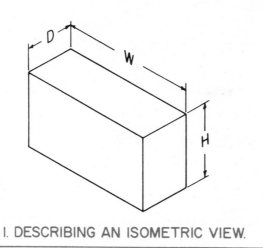

I. DESCRIBING AN ISOMETRIC VIEW.

2. LOCATE CENTER OF DRAWING AREA.

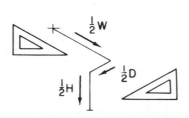

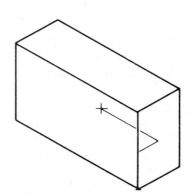

3. PLOT STARTING POINT OF DRAWING.

4. COMPLETE DRAWING.

Fig. 11-19. Above. An easy way to center isometric drawings.
Below. An easy way to center oblique drawings.

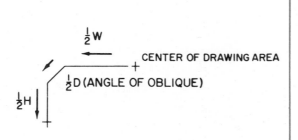

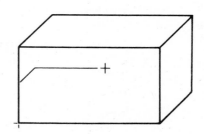

I. PLOT STARTING POINT OF DRAWING.

2. COMPLETE DRAWING.

c. _____.

d. _____.

4. Isometric pictorial drawings are drawn about three base lines. Make a sketch of these base lines.

5. The lines that represent angles in an isometric drawing are known as _____ lines.

6. Prepare a sketch that shows the difference between a cavalier and a cabinet pictorial drawing.

7. Oblique drawings may be drawn at any angle, but _____ angles are usually used.

8. What is an "exploded assembly drawing," and why is it easy to read and understand?

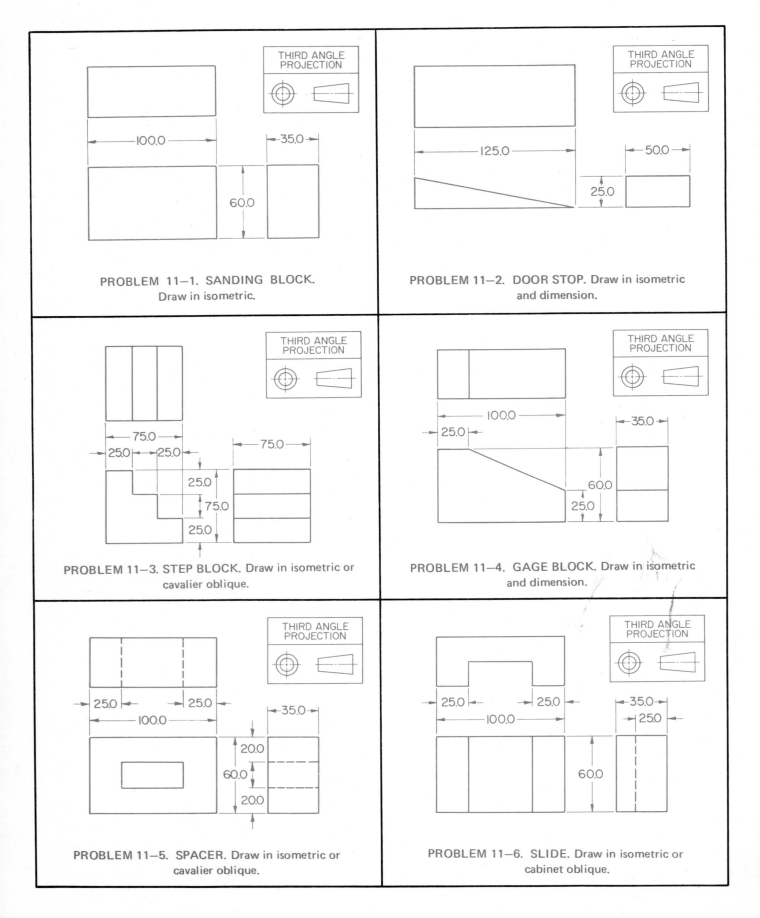

THIRD ANGLE PROJECTION

100.0   35.0

60.0

PROBLEM 11—1. SANDING BLOCK.
Draw in isometric.

THIRD ANGLE PROJECTION

125.0   50.0

25.0

PROBLEM 11—2. DOOR STOP. Draw in isometric and dimension.

THIRD ANGLE PROJECTION

75.0   75.0

25.0   25.0

25.0

75.0

25.0

PROBLEM 11—3. STEP BLOCK. Draw in isometric or cavalier oblique.

THIRD ANGLE PROJECTION

100.0   35.0

25.0

60.0

25.0

PROBLEM 11—4. GAGE BLOCK. Draw in isometric and dimension.

THIRD ANGLE PROJECTION

25.0   25.0   35.0

100.0

20.0

60.0

20.0

PROBLEM 11—5. SPACER. Draw in isometric or cavalier oblique.

THIRD ANGLE PROJECTION

25.0   25.0   35.0

100.0   25.0

60.0

PROBLEM 11—6. SLIDE. Draw in isometric or cabinet oblique.

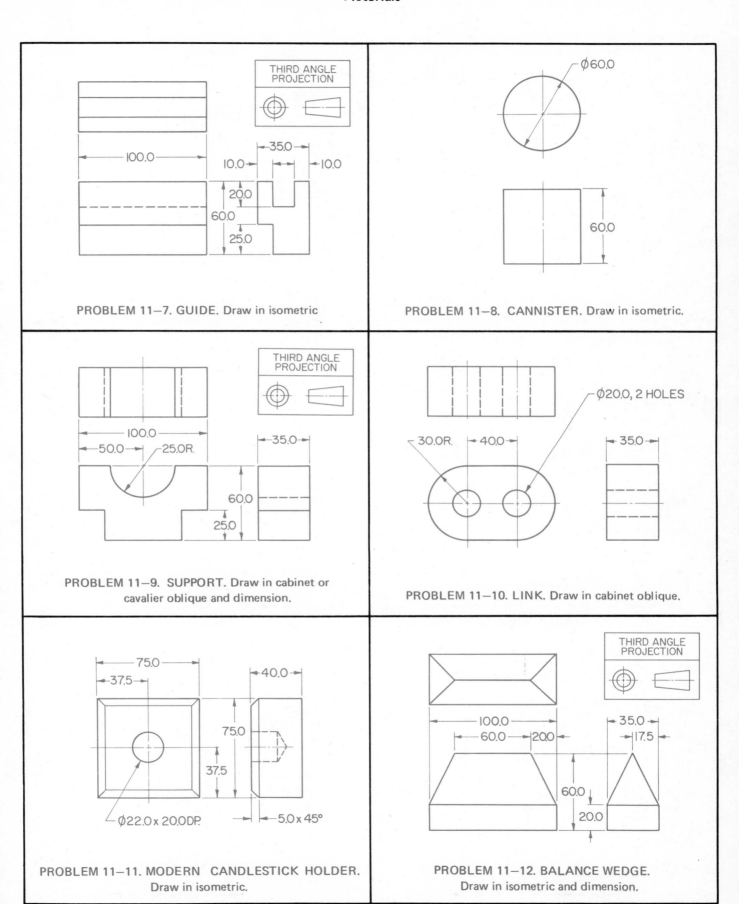

PROBLEM 11—7. GUIDE. Draw in isometric

PROBLEM 11—8. CANNISTER. Draw in isometric.

PROBLEM 11—9. SUPPORT. Draw in cabinet or cavalier oblique and dimension.

PROBLEM 11—10. LINK. Draw in cabinet oblique.

PROBLEM 11—11. MODERN CANDLESTICK HOLDER. Draw in isometric.

PROBLEM 11—12. BALANCE WEDGE. Draw in isometric and dimension.

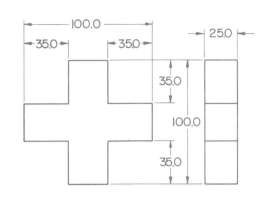

PROBLEM 11–13. CROSS. Draw in one-point perspective.

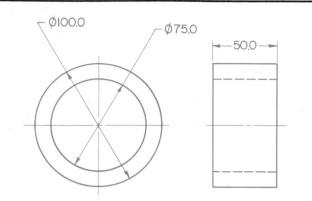

PROBLEM 11–14. BEARING. Draw in isometric or one-point perspective.

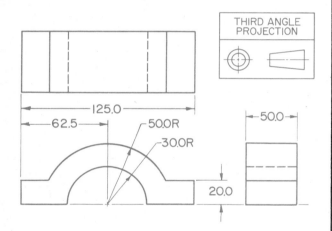

THIRD ANGLE PROJECTION

PROBLEM 11–15. BEARING CAP. Draw in isometric and dimension.

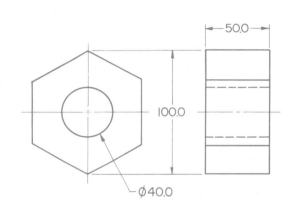

PROBLEM 11–16. HEXAGONAL BASE. Draw in isometric or cabinet oblique.

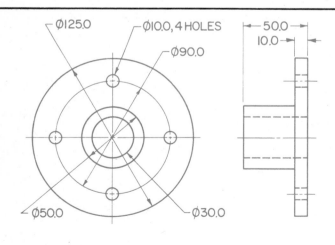

PROBLEM 11–17. FACE PLATE. Draw in isometric.

PROBLEMS:
DESIGN AND DRAW PICTORIALLY
1. BOOK RACK
2. SHOE SHINE BOX
3. MODERN BIRD HOUSE
4. COFFEE TABLE
5. END TABLE
DESIGN AND DRAW AS EXPLODED DRAWING
1. PICTURE FRAME
2. BOOK CASE
3. JEWELRY BOX
4. WALL STORAGE CABINET
5. DESK

PROBLEM 11–18. SPECIAL DESIGN PROBLEMS.

Industry drawing shows size of a nose landing gear used on a modern airplane.   (Bendix)

Unit 12
PATTERN DEVELOPMENT

A pattern is a full-size drawing of the various surfaces of an object stretched out on a flat plane. See Fig. 12-1. A pattern is frequently called a STRETCHOUT.

Patterns or stretchouts are produced by utilizing a form of drafting called PATTERN DEVELOPMENT. The method is also known as SURFACE DEVELOPMENT and SHEET METAL DRAFTING.

Pattern development is important to many occupations and hobbies that require folding or rolling or sheet materials.

Patterns were required to make the clothing and shoes you wear. Pattern development plays an important part in the fabrication of sheet metal ducts and pipes needed in the installation of heating and air conditioning units. Stoves and refrigerators are fabricated from many sheet metal parts. Accurate patterns had to be developed for the parts before the appliance could be put into production.

Drafters in the aerospace industry must be familiar with pattern development techniques. Manufacturing sheet metal parts for airplanes and rockets, Fig. 12-2, involves the use of many patterns.

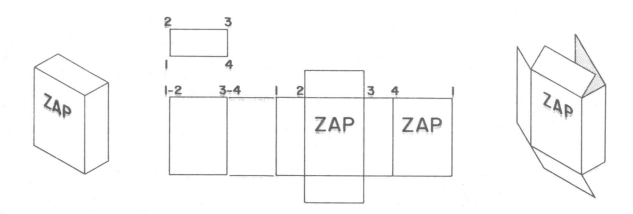

Fig. 12-1. The pattern or stretchout of a cereal type box. Check this layout against a cereal box you have cut open.

Fig. 12-2. Drafters in the aviation industry must be familiar with pattern development. (Gates—Learjet)

If you have ever built a flying model airplane, Fig. 12-3, you know that many patterns are needed to cut the parts to shape for assembly.

Even wallets and handbags were cut to shape using patterns as guides.

As you can see by the wide variety of examples, patterns play an important part in the manufacture of many items. How many more manufactured products can you name where patterns are utilized?

HOW TO DRAW PATTERNS

Regular drafting techniques are used to draw patterns. Pattern development falls into two categories:

PARALLEL LINE DEVELOPMENT. The technique used to make patterns for prisms and cylinders.

RADIAL LINE DEVELOPMENT. A method developed for making patterns for regular tapering forms (cones, pyramids, etc.).

Fig. 12-3. Patterns for wing ribs, fuselage formers and tail sections must be developed before a model airplane can be constructed.

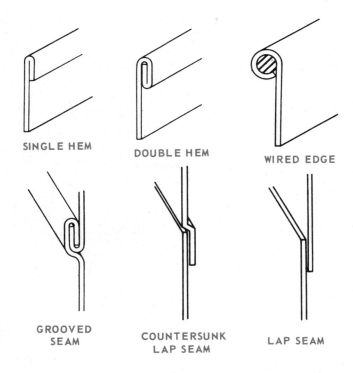

SINGLE HEM   DOUBLE HEM   WIRED EDGE

GROOVED SEAM   COUNTERSUNK LAP SEAM   LAP SEAM

Fig. 12-4. Typical hems, edges and seams used to join and give rigidity to sheet metal. Extra material is required.

Combinations and variations of the basic developments are used to draw patterns for more complex geometric shapes.

While regular drafting techniques are used in pattern development, some lines have additional meanings. Sharp folds or bends are indicated on the stretchout by a visible object line (see PATTERN DEVELOPMENT OF A REGULAR PRISM, page for an example of this type).

Curved surfaces are shown on the pattern by construction lines or center lines (see PATTERN DEVELOPMENT OF A CYLINDER, page 152).

Stretchouts are seldom dimensioned.

When developing a pattern or stretchout, allow additional metal for HEMS, EDGES and SEAMS, Fig. 12-4.

HEMS are used to strengthen the lips of sheet metal objects. They are made in standard sizes, 4.0 mm, 6.0 mm, 10.0 mm, etc.

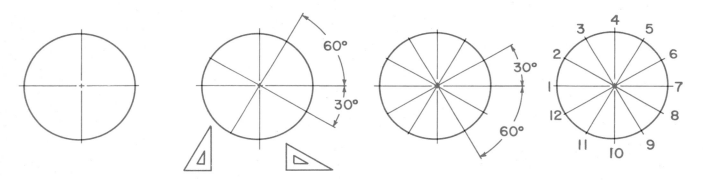

Fig. 12-5. How to divide a circle into twelve (12) equal parts.

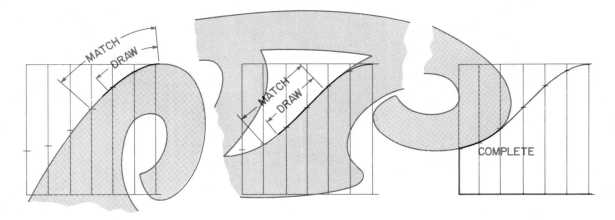

Fig. 12-6. Using a French curve to draw an irregular curve.

The WIRED EDGE gives extra strength and rigidity to sheet metal edges.

SEAMS make is possible to join sheet metal sections. They are usually finished by soldering and/or riveting.

Reference lines and points are needed when developing the stretchout of a circular object. To provide these, the circle is divided into twelve (12) equal parts, Fig. 12-5.

Irregular curves are drawn using a FRENCH CURVE. The points of the irregular curve are plotted. The points may be connected by a lightly sketched line. Match the French curve to the sketched line or points taking care to make the curve of the line flow smoothly. Fig. 12-6 shows how this is done.

TEST YOUR KNOWLEDGE - UNIT 12

(Write answers on a separate sheet of paper.)

1. Pattern development is a form of drafting that ____ _____.

2. Pattern development is also known as:
   a. _____.
   b. _____.

3. Patterns are also known as _____.

4. List five uses of patterns.

5. In pattern development, a heavy solid line (visible object line) indicates that _____ _____.

6. Very light lines (construction lines) and center lines indicate that _____.

7. Make a sketch of a wired edge, a lap seam and a grooved seam.

8. Irregular curves are drawn with a _____.

PATTERN DEVELOPMENT
OF A CYLINDER

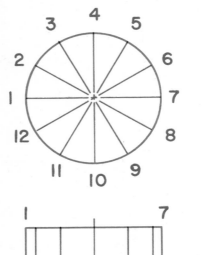

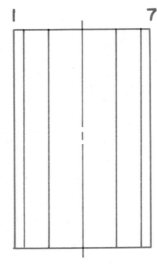

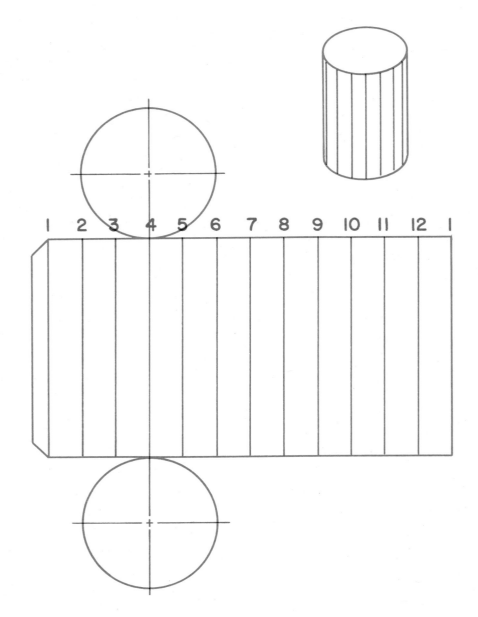

1. Draw front and top views of required cylinder. Divide top view into twelve (12) equal parts and number as shown.
2. The height of the pattern or stretchout is the same as the height of the front view. Project construction lines from the top and bottom of the front view.
3. Allow sufficient space (25.0 mm is adequate) between the front view and the pattern, and draw a vertical line. This will locate line 1 of the pattern.
4. Set your compass or divider from 1 to 2 (the points where the division lines intersect the circle) on the top view. Transfer this distance to the extended lines of the pattern to locate reference lines 1, 2, 3, - 12, 1.
5. Draw the top and bottom tangent to the extended lines.
6. Allow 6.0 mm for seams, and go over all outlines with visible object lines. The lines that represent the curves or circular lines are drawn in color, or they are left as construction lines.

PATTERN DEVELOPMENT OF A
RECTANGULAR PRISM

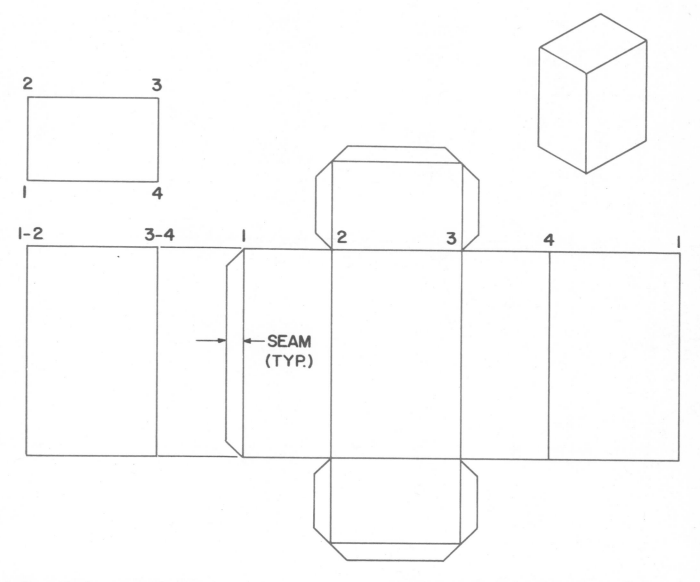

1. Draw the front and top views.
2. The height of the pattern is the same as the height of the front view. Project construction lines from the top and bottom of the front view.
3. Measure over 25.0 mm from the front view and draw a vertical line between the extended lines to locate line 1.
4. Set your compass or divider from 1 to 2 on the top view, and transfer this distance to the extended lines. Locate the other distances in the same manner.
5. Construct the top and bottom as shown.
6. Allow 6.0 mm for seams, and go over all outlines and folds with visible object lines. The pattern may be cut out, folded to shape and cemented together using rubber cement.

PATTERN DEVELOPMENT OF A
TRUNCATED PRISM

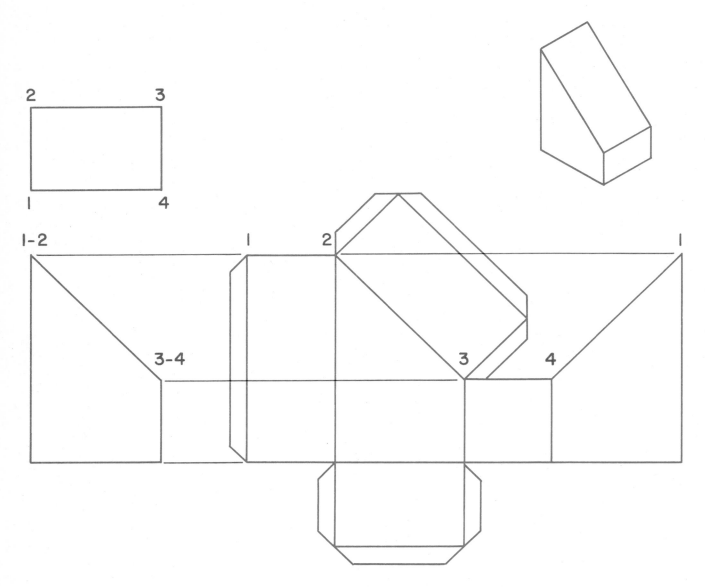

1. Draw front and top views. Number the points as shown.
2. Proceed as in previous examples of pattern development.
3. Mark off and number the folding points. Project point 1 on the front view to line 1 of the stretchout. Repeat with points 2, 3 and 4 to lines 2, 3 and 4.
4. Connect the points 1 to 2, 2 to 3, 3 to 4 and 4 to 1.
5. Draw the top and bottom in position.
6. Allow material for seams, and go over the outline and fold lines with visible object lines.

PATTERN DEVELOPMENT OF A
TRUNCATED CYLINDER

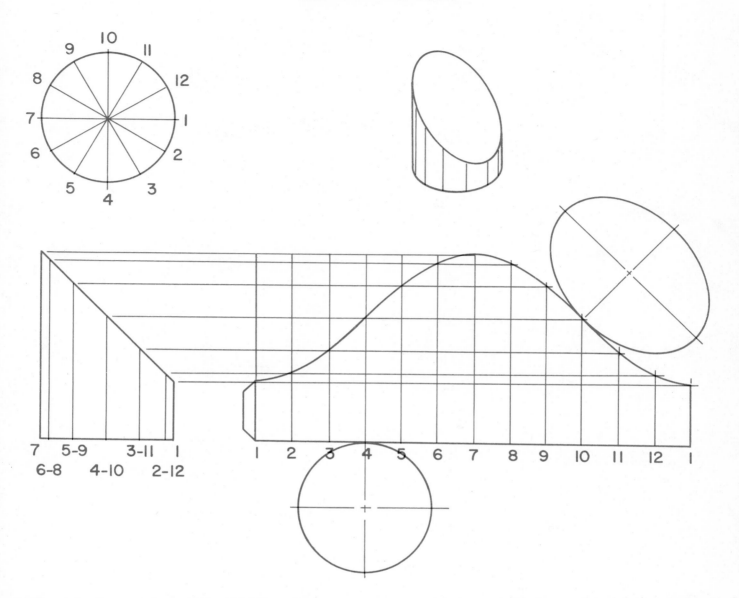

1. Draw the front and top views. Divide and number as shown.
2. Extend lines from the top and bottom of the front view.
3. Allow 25.0 mm between the front view and the pattern and draw line 1.
4. Set your compass or dividers from 1 to 2 on the top view and step off twelve (12) equal divisions on the extended lines of the pattern. Number them as shown in the pattern development.
5. Draw vertical construction lines at each of the above divisions.
6. The curve of the pattern is developed by projecting lines from the points on the front view. Point 1 is projected over until it intersects lines 1 on the pattern; points 2-12 intersect lines 2 and 12; etc. When all of the points are located, connect them with a curved line drawn with a French curve.
7. Complete by adding the top and bottom. The top is developed as an auxiliary view.

PATTERN DEVELOPMENT
OF A PYRAMID

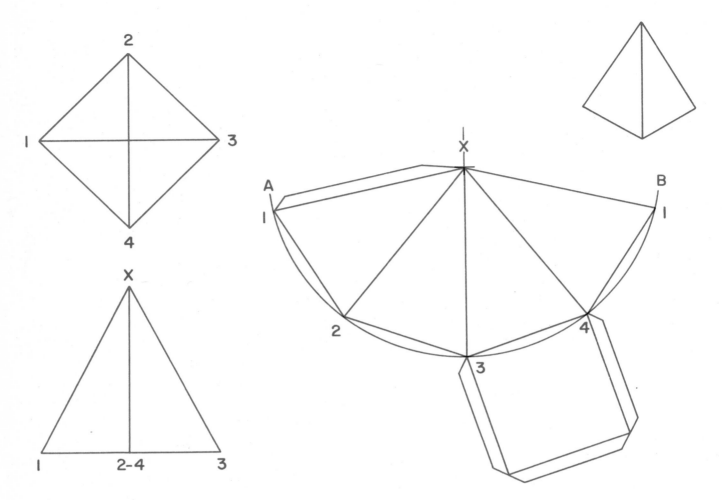

1. Draw the front and top views. Number as shown.
2. Locate center line X of the stretchout.
3. Set your compass to a radius equal to X-1 on the front view and, using the above center, draw arc A—B.
4. Draw a vertical line through center X and arc A—B.

5. Set your compass from 1 to 2 on the top view and at the point where the vertical line intersects the arc as the starting point, step off two (2) divisions on each side of the line. (Points 1-2-3-4-1 on the stretchout.)
6. Connect the points, draw the bottom in place, and go over the outline and folds with object lines.

PATTERN DEVELOPMENT
OF A CONE

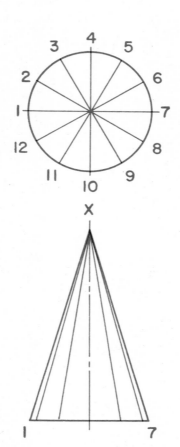

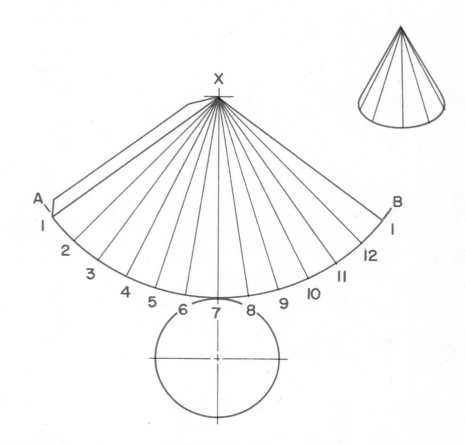

1. Draw the front and top views. Divide the top view into twelve (12) equal parts. Number as shown.
2. Locate center line X of the stretchout.
3. Set your compass from X to 1 on the front view and, with X of the stretchout as the center, draw arc A—B.
4. Draw a vertical construction line through center line X and the arc.
5. Set your compass from 1 to 2 on the top view and, with the point where the vertical line intersects the arc as the starting point, step off six (6) divisions on each side of the line. (Points 1-2-3-4 etc. on the pattern.)
6. Go over the outline carefully with object lines.
7. The lines that represent the curved portion may be drawn in color or left as construction lines.

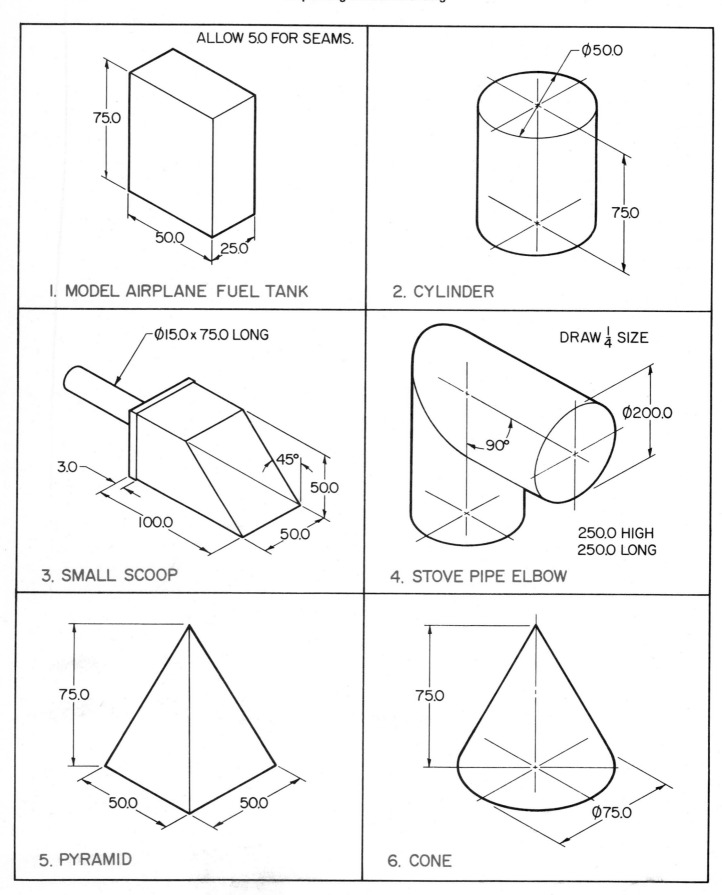

ALLOW 5.0 FOR SEAMS.

75.0

50.0 25.0

1. MODEL AIRPLANE FUEL TANK

Ø50.0

75.0

2. CYLINDER

Ø15.0 x 75.0 LONG

3.0

100.0

45°

50.0

50.0

3. SMALL SCOOP

DRAW ¼ SIZE

90°

Ø200.0

250.0 HIGH
250.0 LONG

4. STOVE PIPE ELBOW

75.0

50.0 50.0

5. PYRAMID

75.0

Ø75.0

6. CONE

PROBLEMS 12—1 to 12—6. Develop patterns for the objects above. Use the dimensions provided in the problems.

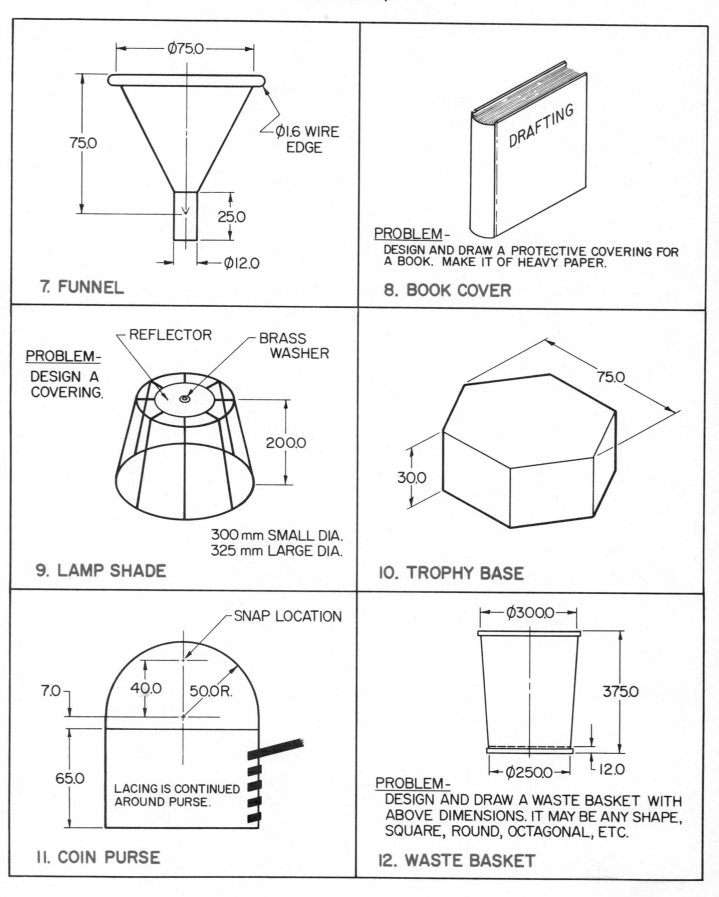

Ø75.0

75.0

Ø1.6 WIRE EDGE

25.0

Ø12.0

7. FUNNEL

DRAFTING

PROBLEM—
DESIGN AND DRAW A PROTECTIVE COVERING FOR A BOOK. MAKE IT OF HEAVY PAPER.

8. BOOK COVER

REFLECTOR — BRASS WASHER

PROBLEM—
DESIGN A COVERING.

200.0

300 mm SMALL DIA.
325 mm LARGE DIA.

9. LAMP SHADE

75.0

30.0

10. TROPHY BASE

SNAP LOCATION

7.0

40.0   50.0R.

65.0

LACING IS CONTINUED AROUND PURSE.

11. COIN PURSE

Ø300.0

375.0

Ø250.0   12.0

PROBLEM—
DESIGN AND DRAW A WASTE BASKET WITH ABOVE DIMENSIONS. IT MAY BE ANY SHAPE, SQUARE, ROUND, OCTAGONAL, ETC.

12. WASTE BASKET

PROBLEMS 12—7 to 12—12. Develop patterns for the objects above. Use the dimensions provided in the problems.

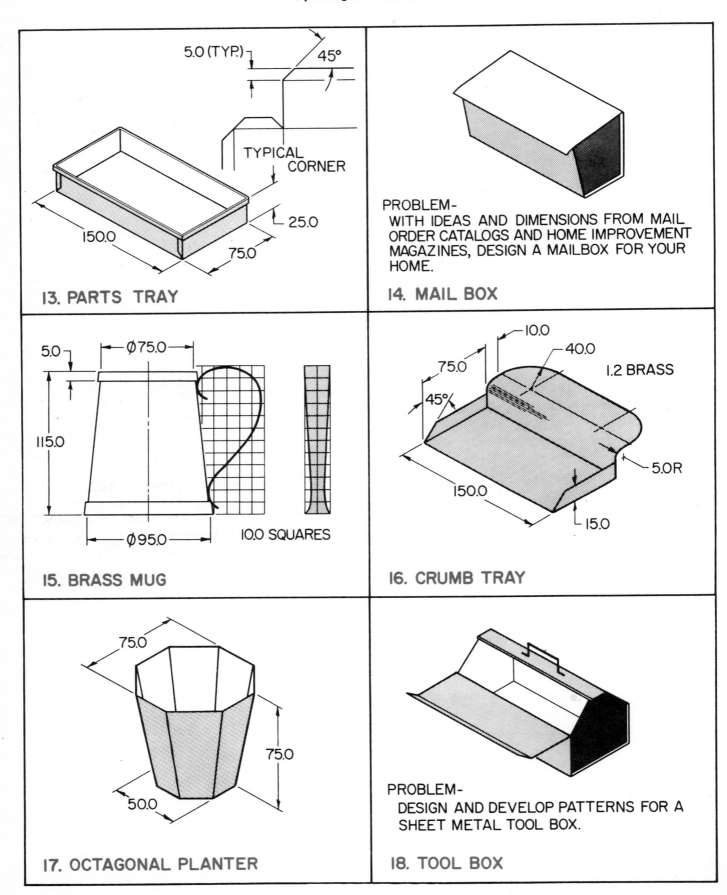

5.0 (TYP.) 45°

TYPICAL CORNER

25.0

150.0

75.0

13. PARTS TRAY

PROBLEM-
WITH IDEAS AND DIMENSIONS FROM MAIL ORDER CATALOGS AND HOME IMPROVEMENT MAGAZINES, DESIGN A MAILBOX FOR YOUR HOME.

14. MAIL BOX

5.0 Ø75.0

115.0

Ø95.0

10.0 SQUARES

15. BRASS MUG

10.0 40.0

75.0 1.2 BRASS

45°

150.0 5.0R

15.0

16. CRUMB TRAY

75.0

75.0

50.0

17. OCTAGONAL PLANTER

PROBLEM-
DESIGN AND DEVELOP PATTERNS FOR A SHEET METAL TOOL BOX.

18. TOOL BOX

PROBLEMS 12–13 to 12–18. Develop patterns for the objects above. Use the dimensions provided in the problems.

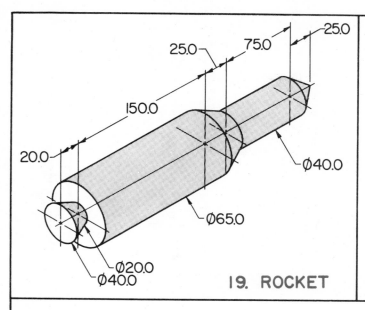

19. ROCKET

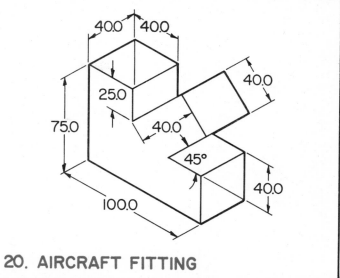

20. AIRCRAFT FITTING

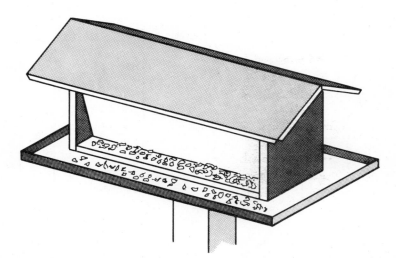

PROBLEM-

DESIGN AND DEVELOP PATTERNS FOR A BIRD FEEDER. THE ROOF MUST BE HINGED FOR EASY LOADING OF FEED. CONSTRUCT A CARDBOARD MODEL OF YOUR DESIGN.

21. BIRD FEEDER

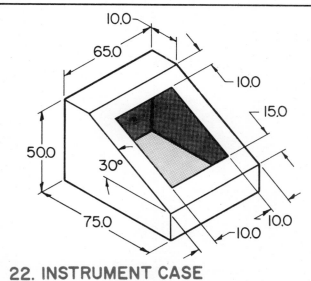

22. INSTRUMENT CASE

DESIGN PROBLEMS-

DESIGN AND DEVELOP PATTERNS FOR THE FOLLOWING PROJECTS. MAKE A CARDBOARD MODEL OF YOUR CHOICE.

 - PLANTER
 - DUST PAN
 - CHARCOAL SCOOP
 - BOOK ENDS
 - LEATHER HAND BAG
 - JEWEL BOX
 - BRASS COIN BANK

23. PROBLEMS

PROBLEMS 12—19 to 12—23. Develop patterns for the objects above. Use the dimensions provided in the problems.

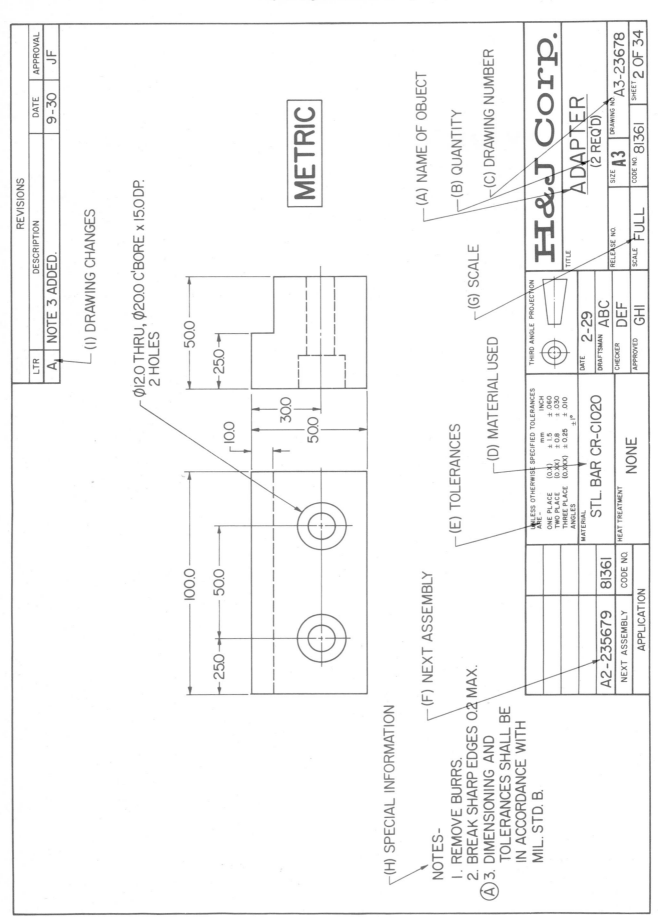

Fig. 13-1. A working drawing.

Unit 13
WORKING DRAWINGS

It would not be practical for industry to manufacture a product without using drawings which provide complete manufacturing details.

Drawings required range from a single freehand sketch for a simple job to many thousands of drawings needed to manufacture a complex product like the helicopter shown in Fig. 1-1, page 7. The drawings which tell the craftworker what to make and establish the standards to which the work must be done are called WORKING DRAWINGS, Fig. 13-1.

Working drawings fall into two categories, DETAIL DRAWINGS and ASSEMBLY DRAWINGS. An example of a DETAIL DRAWING is given in Fig. 13-2. It includes a view (or views) of the product, with dimensions and other pertinent information required to make the part. See Fig. 13-1.

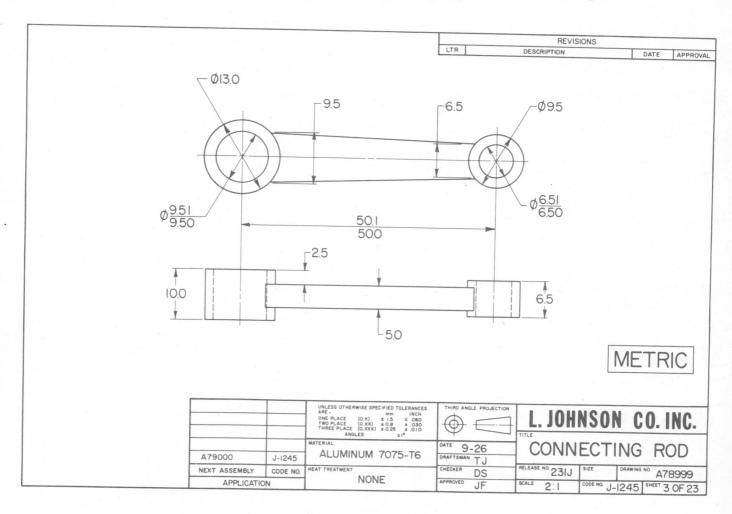

Fig. 13-2. A detail drawing.

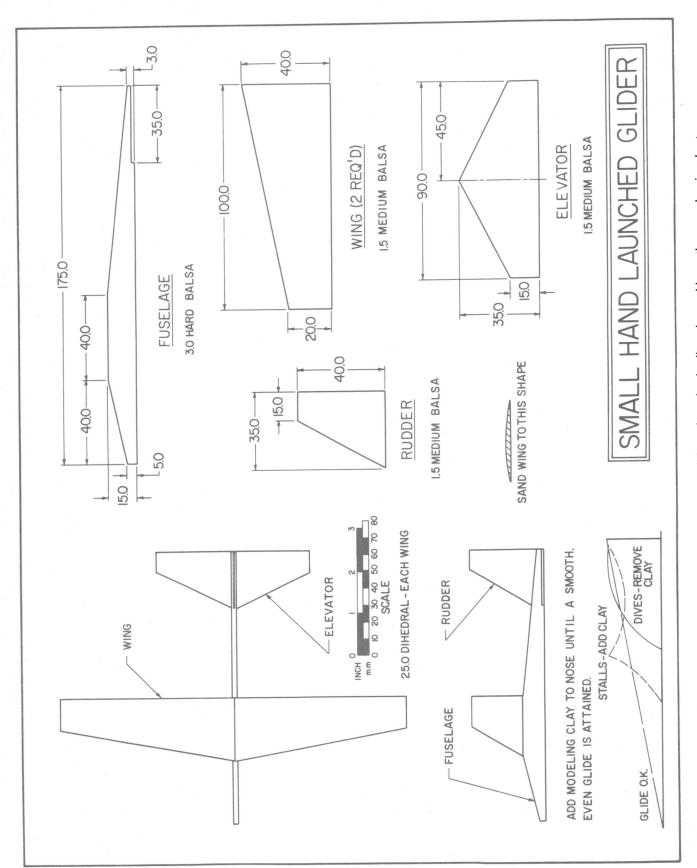

Fig. 13-3. If the item is small enough, it is permissible to draw the details and assembly on the same drawing sheet.

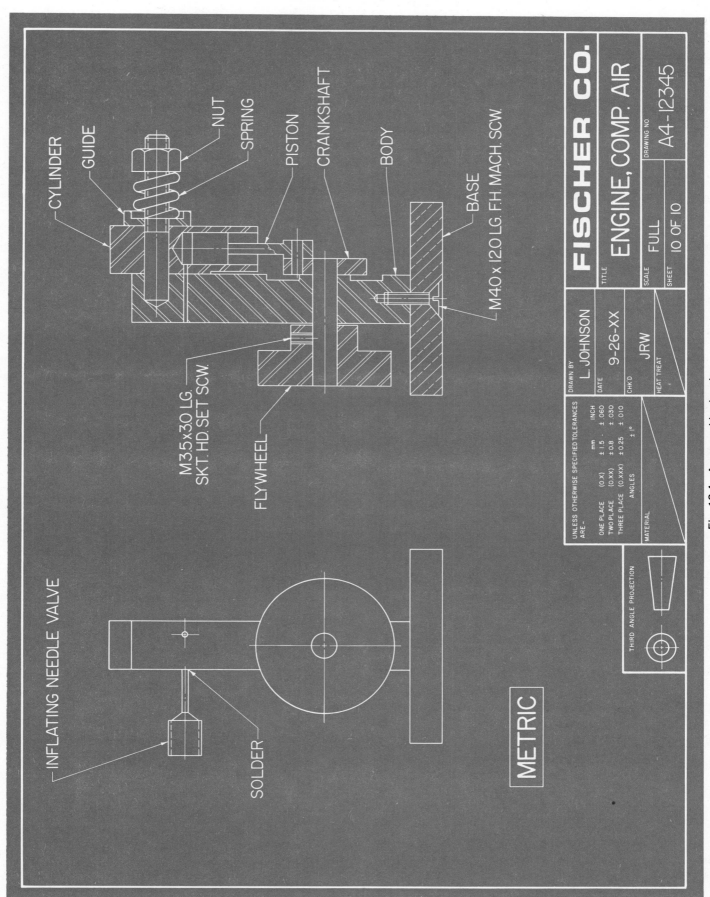

Fig. 13-4. An assembly drawing.

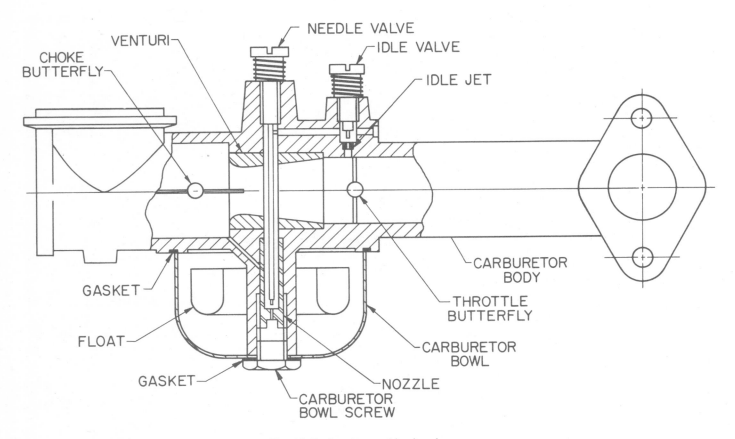

Fig. 13-5. A subassembly drawing.

1. NAME OF THE OBJECT. The name assigned to the part should appear on all drawings making reference to the piece.
2. QUANTITY. The number of units needed in each assembly.
3. DRAWING NUMBER. Number assigned to the drawing for filing and reference.
4. MATERIAL USED. The exact material specified to make the part.
5. TOLERANCES. The permissible deviation, either oversize or undersize, from a basic dimension.
6. NEXT ASSEMBLY. The name given to the major assembly on which the part is to be used.
7. SCALE. Drawings made other than full size are called scale drawings.
8. SPECIAL INFORMATION. Information pertinent to the correct manufacture of the part that is not included on the various views of the object.
9. REVISIONS. Changes that have been made on the original drawing.

In most instances, the detail drawing provides information on a single part. However, it is permissible to draw all of the parts and the assembly of a small or simple mechanism on the same sheet. Note the use of this technique in Fig. 13-3.

An example of the second type of working drawing, an ASSEMBLY DRAWING, is given in Fig. 13-4. As its name implies, views show where and how the various parts fit into the assembled product. They also show how the complete object will look. When the product is large or complex, it may not be possible to present all of the information one one sheet. A SUBASSEMBLY DRAWING, Fig. 13-5, is utilized when this problem arises. This type of drawing illustrates the assembly of only a small portion of the complete object.

Under certain situations, a drawing may include a PARTS LIST, Fig. 13-6, and/or a BILL OF MATE-

PARTS LIST		
NO.	NAME	QUAN.
1	CRANKCASE	1
2	CRANKSHAFT	1
3	CRANKCASE COVER	1
4	CYLINDER	2
5	PISTON	2

Fig. 13-6. Parts list.

RIALS, Fig. 13-7. The information on the parts list is read from the top down when the list is located at the upper right corner of the drawing sheet, or, read from the bottom up when located above the title block. Either position is acceptable. The parts usually are listed according to size or in their order of importance.

An IDENTIFYING NUMBER, Fig. 13-8, is placed on each sheet for convenience in filing or rapid location of the drawing.

AI776	NUT, HEX.	BRASS	6
AI985	BOLT, SPEC.	BRASS	6
BI766	PLATE	STEEL	2
BI767	CYLINDER	CAST IRON	2
PT. NO.	NAME	MATERIAL	QUAN.

BILL OF MATERIALS

Fig. 13-7. Bill of materials.

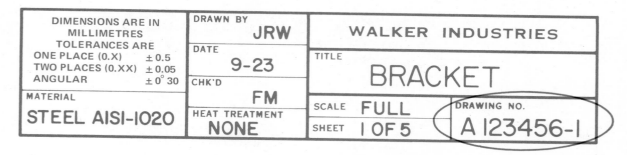

Fig. 13-8. For convenience in filing and locating drawings, industry gives each plate an identifying number.

TEST YOUR KNOWLEDGE - UNIT 13

(Write answers on a separate sheet of paper.)

1. Why are working drawings used?
2. The _____ drawing has a picture of the part with dimensions and other information needed to make the object.
3. The _____ drawing shows where and how the object described in the above drawing fits into the complete assembly of the object.
4. On large or complex objects, _____ _____ drawings frequently are used. These drawings show the assembly of only a small portion of the complete object.

The drawing described in Question No. 2 includes much information necessary to manufacture a part. Some of this information is described below. Place the letter of the correct description in the blank before the words below.

5. _____ QUANTITY
6. _____ TOLERANCES
7. _____ MATERIAL USED
8. _____ SCALE
9. _____ REVISIONS

A. The exact material specified to make the part.
B. Drawings make other than full size are called this.
C. The number of units needed in each assembly.
D. The permissible deviation, either oversize or undersize, from a basic dimension.
E. Changes that have been made on the original drawing.

OUTSIDE ACTIVITIES

1. From a local industry, try to obtain samples of detail drawings.
2. Try to obtain samples of assembly drawings.
3. Try to obtain samples of subassembly drawings.
4. Prepare a bulletin board display using the theme WORKING DRAWINGS.
5. Design a suitable title block for the drawing sheets of a company that you are planning to start.
6. Secure the piston and rod assembly from a small, single cylinder gasoline engine and prepare detail drawings for them. Indicate all dimensions as decimals. It may be necessary for you to learn how to read a micrometer to make accurate measurements.
7. Design a foot stool and prepare the drawings necessary to mass-produce it in the school shop. Make your drawings on tracing vellum so that a number of prints can be made.
8. Design a coffee table or end table. Prepare the drawings needed to make it in the school shop.

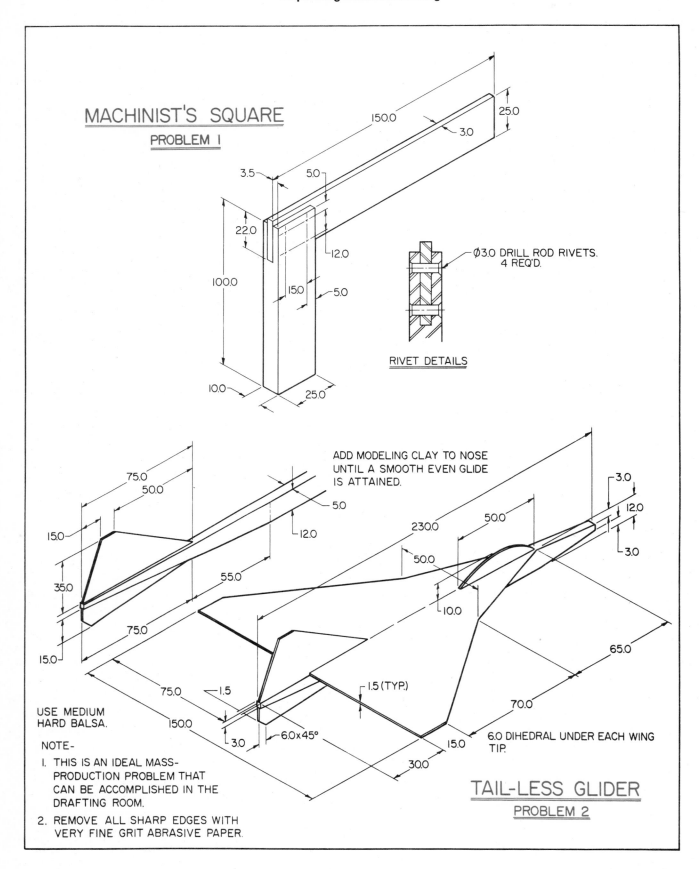

MACHINIST'S SQUARE
PROBLEM 1

150.0

25.0

3.0

3.5

5.0

22.0

12.0

Ø3.0 DRILL ROD RIVETS.
4 REQ'D.

100.0

15.0

5.0

RIVET DETAILS

10.0

25.0

ADD MODELING CLAY TO NOSE
UNTIL A SMOOTH EVEN GLIDE
IS ATTAINED.

75.0

50.0

5.0

3.0

12.0

15.0

12.0

230.0

50.0

3.0

35.0

55.0

50.0

65.0

15.0

10.0

75.0

USE MEDIUM
HARD BALSA.

75.0

1.5

1.5 (TYP.)

70.0

NOTE—

150.0

3.0

6.0×45°

15.0

6.0 DIHEDRAL UNDER EACH WING
TIP.

1. THIS IS AN IDEAL MASS-
PRODUCTION PROBLEM THAT
CAN BE ACCOMPLISHED IN THE
DRAFTING ROOM.

30.0

TAIL-LESS GLIDER
PROBLEM 2

2. REMOVE ALL SHARP EDGES WITH
VERY FINE GRIT ABRASIVE PAPER.

PROBLEM 13—1. Above. MACHINIST'S SQUARE. Make detail and assembly drawings. PROBLEM 13—2. Below.
TAIL-LESS GLIDER. Prepare detail and assembly drawings. Design the necessary. patterns, fixtures, etc., needed to
mass-produce the glider.

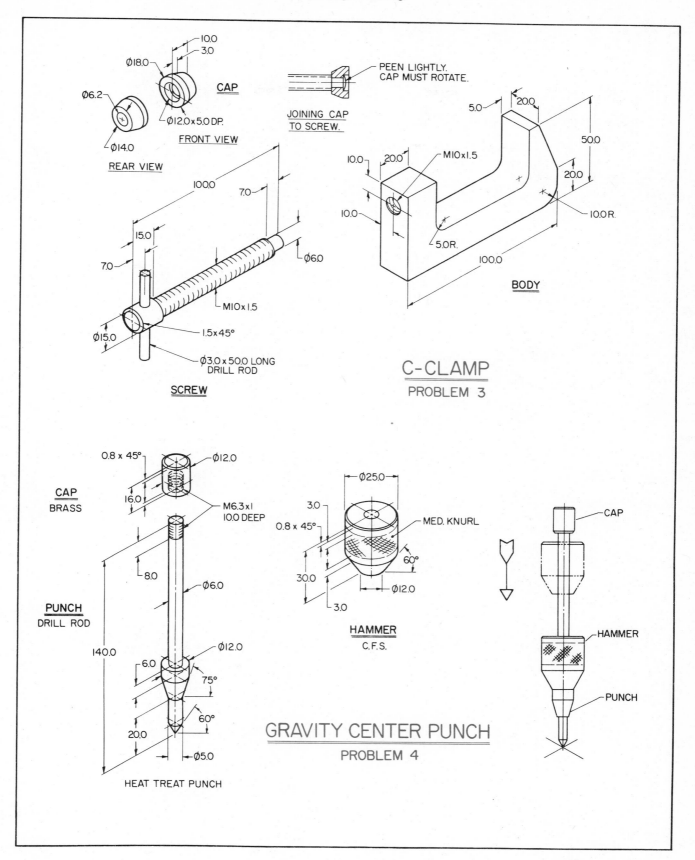

10.0
3.0
Ø18.0
Ø6.2
CAP
Ø12.0 x 5.0 DP.
Ø14.0
FRONT VIEW
REAR VIEW

PEEN LIGHTLY.
CAP MUST ROTATE.
**JOINING CAP
TO SCREW.**

100.0
7.0
15.0
7.0
Ø6.0
M10 x 1.5
Ø15.0
1.5 x 45°
Ø3.0 x 50.0 LONG
DRILL ROD
SCREW

5.0
20.0
50.0
10.0
20.0
M10 x 1.5
20.0
10.0R.
10.0
5.0 R.
100.0
BODY

C-CLAMP
PROBLEM 3

0.8 x 45°
Ø12.0
CAP
BRASS
16.0
M6.3 x 1
10.0 DEEP

PUNCH
DRILL ROD
8.0
Ø6.0
140.0
Ø12.0
6.0
75°
60°
20.0
Ø5.0
HEAT TREAT PUNCH

Ø25.0
3.0
0.8 x 45°
MED. KNURL
60°
30.0
Ø12.0
3.0
HAMMER
C.F.S.

CAP
HAMMER
PUNCH

GRAVITY CENTER PUNCH
PROBLEM 4

PROBLEM 13—3. Above. C—CLAMP. Prepare detail and assembly drawings. PROBLEM 13—4. Below. GRAVITY CENTER PUNCH. Make detail drawings.

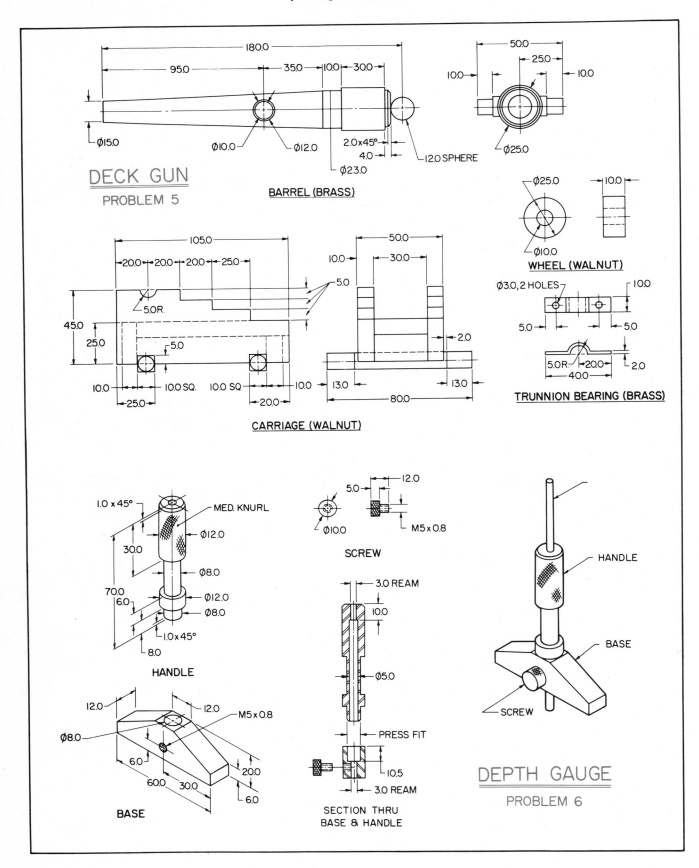

DECK GUN

PROBLEM 5

BARREL (BRASS)

WHEEL (WALNUT)

TRUNNION BEARING (BRASS)

CARRIAGE (WALNUT)

HANDLE

SCREW

BASE

SECTION THRU
BASE & HANDLE

DEPTH GAUGE

PROBLEM 6

PROBLEM 13—5. Above. DECK GUN. Prepare an assembly drawing. PROBLEM 13—6. Below. DEPTH GAUGE. Make detail and assembly drawings.

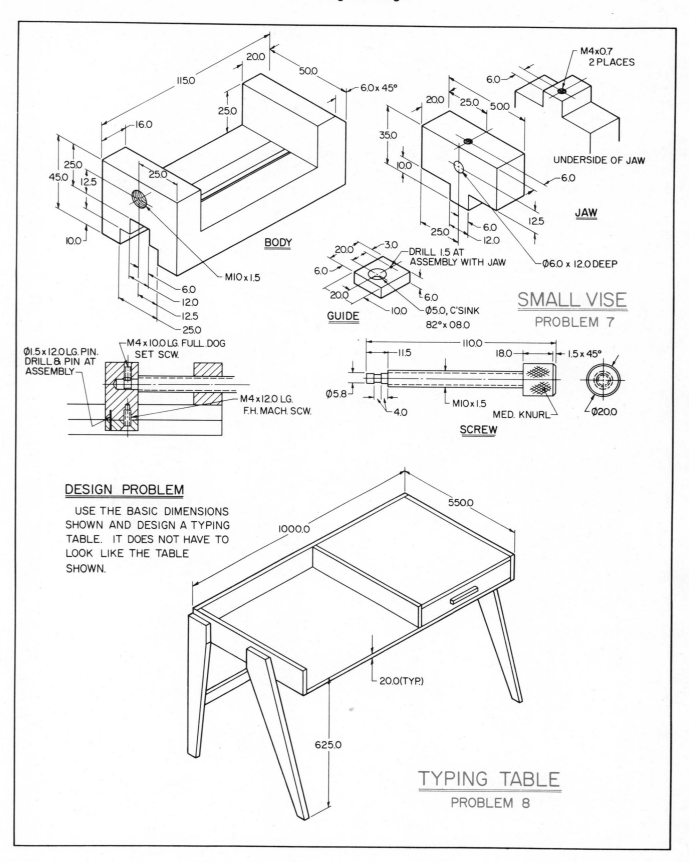

20.0

115.0

50.0

6.0 x 45°

16.0

25.0

25.0

25.0

45.0

12.5

25.0

10.0

M10 x 1.5

6.0
12.0
12.5
25.0

M4x0.7
2 PLACES

6.0

20.0 25.0 50.0

35.0

10.0

UNDERSIDE OF JAW

6.0

JAW

25.0

6.0
12.0

Ø6.0 x 12.0 DEEP

12.5

BODY

Ø1.5 x 12.0 LG. PIN.
DRILL & PIN AT
ASSEMBLY

M4 x 10.0 LG. FULL DOG
SET SCW.

M4 x 12.0 LG.
F.H. MACH. SCW.

20.0   3.0

6.0

20.0

10.0

DRILL 1.5 AT
ASSEMBLY WITH JAW

6.0

Ø5.0, C'SINK
82° x 08.0

GUIDE

SMALL VISE
PROBLEM 7

11.5   110.0   18.0   1.5 x 45°

Ø5.8

4.0

M10 x 1.5

MED. KNURL

Ø20.0

SCREW

DESIGN PROBLEM

USE THE BASIC DIMENSIONS
SHOWN AND DESIGN A TYPING
TABLE. IT DOES NOT HAVE TO
LOOK LIKE THE TABLE
SHOWN.

550.0

1000.0

20.0(TYP.)

625.0

TYPING TABLE
PROBLEM 8

PROBLEM 13–7. Above. SMALL VISE. Prepare detail and assembly drawings. PROBLEM 13–8. Below. TYPING TABLE. Design a table using the dimensions given. Prepare the drawings that will be needed to make your table.

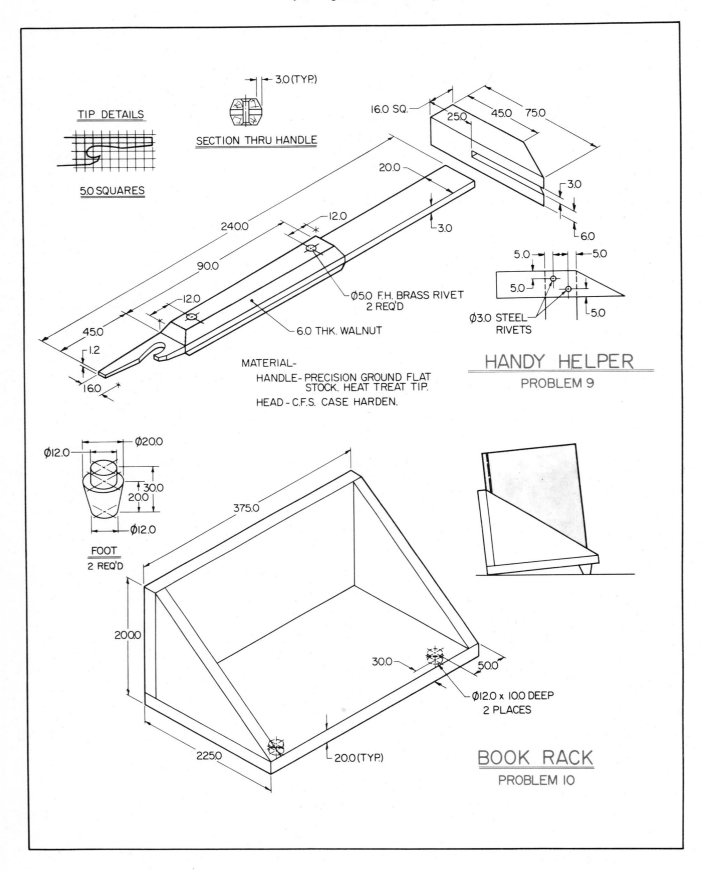

3.0 (TYP.)

TIP DETAILS

SECTION THRU HANDLE

5.0 SQUARES

16.0 SQ.

25.0   45.0   75.0

3.0

6.0

20.0

240.0

12.0

3.0

90.0

12.0

5.0   5.0

5.0

5.0

Ø5.0 F.H. BRASS RIVET
2 REQ'D

Ø3.0 STEEL
RIVETS

45.0

6.0 THK. WALNUT

1.2

16.0

MATERIAL-
HANDLE- PRECISION GROUND FLAT
STOCK. HEAT TREAT TIP.
HEAD - C.F.S. CASE HARDEN.

HANDY HELPER
PROBLEM 9

Ø12.0

Ø20.0

30.0

20.0

Ø12.0

FOOT
2 REQ'D

375.0

200.0

30.0

50.0

Ø12.0 x 10.0 DEEP
2 PLACES

225.0

20.0 (TYP.)

BOOK RACK
PROBLEM 10

PROBLEM 13—9. Above. HANDY HELPER. Prepare detail and assembly drawings. PROBLEM
13—10. Below. BOOK RACK. Make a fully dimensioned assembly drawing.

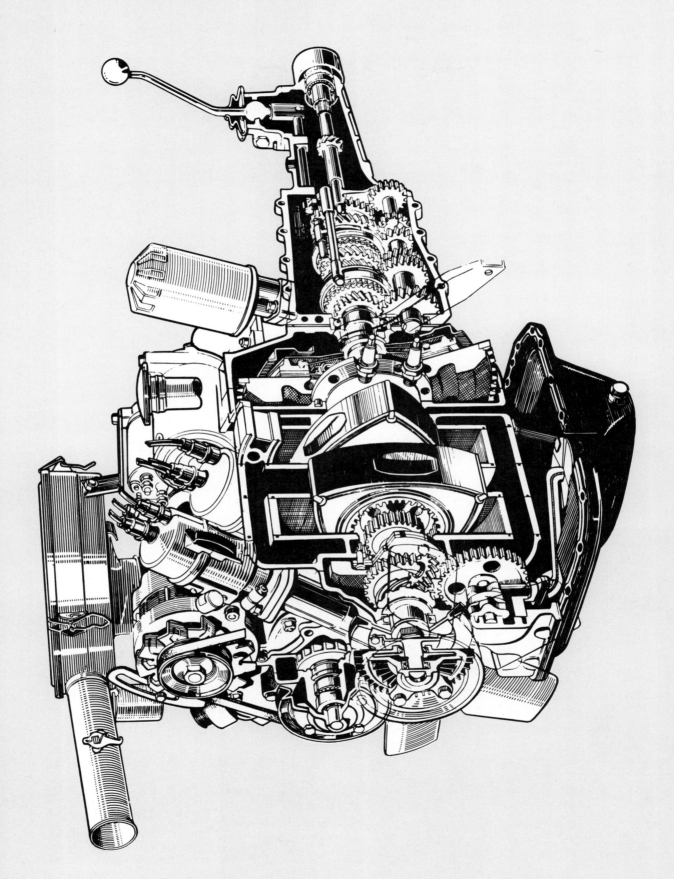

Rotary combustion engines built to metric specifications are used in many fields. RC engines can be built as small as 300 cm^3 per working chamber. (Mazda Motors)

Unit 14
PRINT MAKING

POSITIVE PRINT

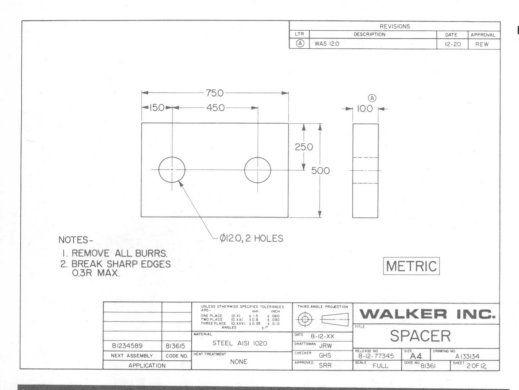

BLUEPRINT

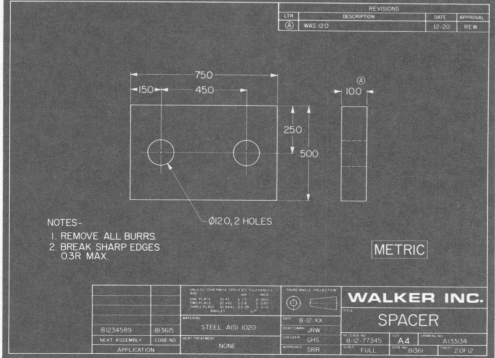

Fig. 14-1. The true blueprint is a print with white lines on a blue background. The term "blueprint" is commonly used when referring to all types of prints regardless of the color of the lines or the background.

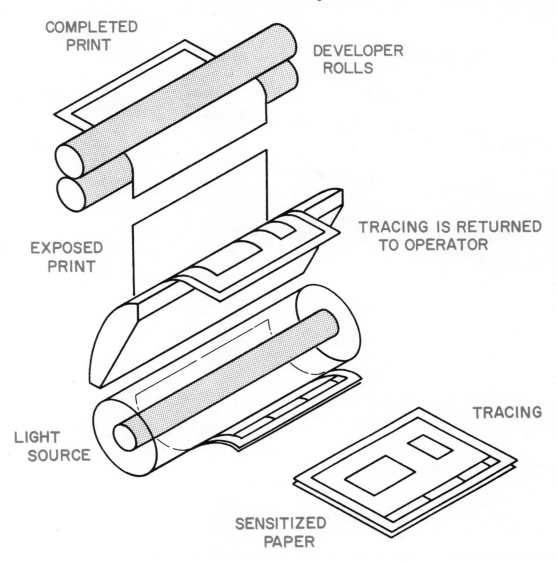

COMPLETED PRINT

DEVELOPER ROLLS

EXPOSED PRINT

TRACING IS RETURNED TO OPERATOR

TRACING

LIGHT SOURCE

SENSITIZED PAPER

Fig. 14-2. The diazo process. On most machines, the print is made in one continuous process.

Original drawings are seldom used on a job because they would soon become worn and soiled and difficult to read. Also, several sets of identical drawings are frequently needed at the same time because craftworkers at different locations are engaged in producing the product or building the structure.

High cost makes it impractical to draw up a set of plans for each person who needs them and to replace plans damaged or ruined.

When several sets of plans are needed, the original drawings are reproduced or duplicated. The technique used must produce accurate copies of the originals, must not destroy the original drawings and must be speedy and reasonable in cost.

Several of the more important duplicating processes will be described in this Unit. The original drawing to be reproduced usually is made on semi-transparent (translucent) material such as tracing paper, cloth or film.

BLUEPRINTS

The BLUEPRINT PROCESS, Fig. 14-1, is the oldest of the techniques employed to duplicate drawings. The print consists of white lines on a blue background.

To make a print by the blueprint method, the tracing is placed in contact with a sensitized paper (unexposed blueprint paper). The two sheets are exposed to a bright light. The light cannot penetrate

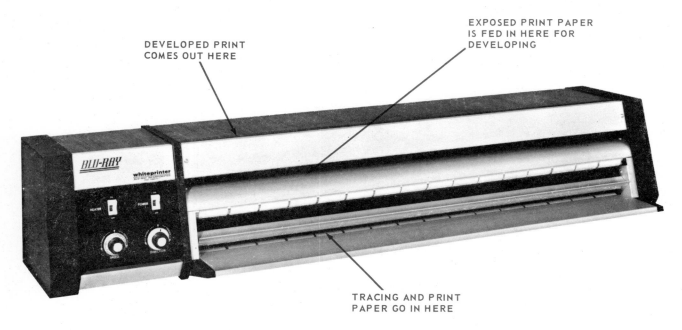

DEVELOPED PRINT
COMES OUT HERE

EXPOSED PRINT PAPER
IS FED IN HERE FOR
DEVELOPING

TRACING AND PRINT
PAPER GO IN HERE

Fig. 14-3. A combination printer and developer type diazo copying machine. (Blue—Ray, Inc.)

the opaque lines of the tracing, and exposure takes place only where the light strikes the sensitive coating on the print paper.

The exposed print is developed by washing in water. The lines are "fixed" by washing in a solution of potassium dichromate crystals dissolved in water to prevent the lines from fading. After a final washing, the print is dried and pressed.

DIAZO PROCESS

The DIAZO PROCESS, Fig. 14-2, is a versatile copying technique for making direct positive copies

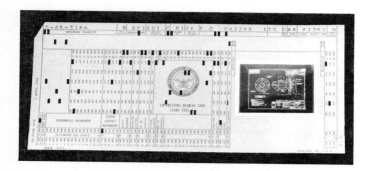

Fig. 14-4. The small negative on the microfilm aperture card is enlarged by a photographic process to the desired print size. The card is easier to store than full size prints and, when needed, can be retrieved by computerized techniques. Some microfilm methods utilize the film in roll form.

(dark lines on light background) of anything that is drawn, written or typed on semitransparent paper, cloth or film. The process produces the copies quickly and is inexpensive.

The print is made by placing the tracing in contact with sensitized copying material (paper, cloth or film) and exposing it to light. After the tracing has been removed, the exposed copy is developed by passing it through ammonia vapors.

With most modern diazo copying machines, Fig. 14-3, the print is made in one continuous process. In addition to turning out the finished copy in seconds, it is clean, odorless and quiet.

Using this process, black-on-white, black-on-color, color-on-white or color-on-color prints can be produced.

THE MICROFILM PROCESS

The MICROFILM PROCESS, Fig. 14-4, was originally designed to reduce storage facilities and to protect prints from loss. With this process, the drawing is reduced by photographic means. Finished negatives can be stored in roll form, or on cards adaptable to computerized storage and retrieval. To produce a working print, the microfilm image is retrieved from the files and enlarged (called BLOW-BACKS) on photographic paper. The print is dis-

Fig. 14-5. A microfilm reader. The enlarged print can be studied on a view screen in this unit and then made into a print of the required size.   (RECORDAK)

carded or destroyed when it is no longer needed. Microfilms can also be viewed on a READER to check details. See Fig. 14-5.

HOW TO MAKE A TRACING

The first step in making a tracing for reproduction is to draw it on a semitransparent or translucent material. There are many types of this material available in sheet or roll form.

TRACING PAPER is frequently used because it is inexpensive. Tracing paper is used extensively for making preliminary plans, sketching and the like, in pencil. They are natural papers made fairly strong and durable, but not very transparent.

TRACING CLOTH is excellent for making original tracings that must withstand severe handling. Changes and alterations can be made without damaging the drawing quality of the cloth.

Tracing cloths are more expensive than papers but retain their drawing and reproduction qualities almost indefinitely. They are made from specially treated, fine grade linen cloth. One surface has a matte finish to take pencil or ink.

TRACING FILMS made from acetate, Mylar or polyester are ideally suited for work requiring a high degree of dimensional stability (it will not expand or contract in heat, cold or high humidity). They have high transparency and make prints of maximum contrast. One surface has a matte finish that will take pencil or ink. Corrections are made with an eraser or damp cloth. Tracing films are tough and tear resistant. They do not discolor or become brittle with age.

PENCIL TRACINGS

Sharp, clear reproductions can be made from pencil tracings with modern reproduction equipment

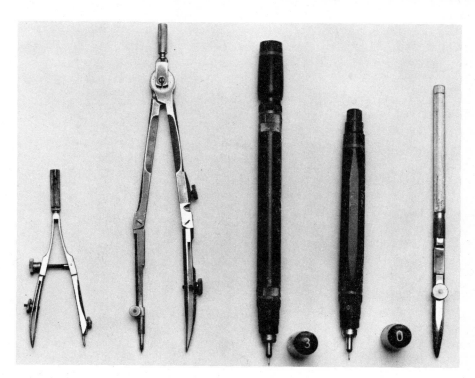

Fig. 14-6. Drafting equipment used in inking: compasses, technical fountain pens and ruling pen.

CORRECTLY INKED LINE

INSTRUMENT ACCIDENTALLY PUSHED
INTO FRESHLY INKED LINE

PEN NOT HELD VERTICALLY

INK RAN UNDER EDGE OF
INSTRUMENT. PEN NOT HELD VERTICAL

NOT ENOUGH INK TO COMPLETE LINE

TOO MUCH INK IN PEN

Fig. 14-7. Common mistakes made in inking.

and materials. At present most drawings used by industry are done in pencil.

Many drafters make original drawings on the tracing material that will be used to produce the copies. For complex jobs, the layout is frequently traced from the original drawing onto other sheets which are used in making copies (this is where we get the term TRACING).

Tracings for reproduction are made the same way as the conventional drawings you have been making. Block in the needed views with a 4H or 6H pencil. If they are drawn lightly enough, they will not have to be erased. Complete the drawing and add notes and dimensions, using a 2H pencil. Keep the lines uniformly sharp and dense as this type line reproduces best.

Every effort must be made to keep your drawings clean. Working with clean hands and a piece of paper under your hand when you letter dimensions and notes will help.

INKED TRACINGS

Inked tracings are more difficult to prepare than tracings in pencil. A knowledge of inking techniques and much patience and practice are required to produce acceptable inked tracings.

A dense black, waterproof ink (called INDIA INK) is used. Lines are drawn with a RULING PEN or a TECHNICAL FOUNTAIN PEN, Fig. 14-6. The ruling pen can be adjusted to make different width lines. When using a technical fountain pen, however, different size points are needed to vary line width.

Hold the pen in much the same manner you would hold a pencil. Practice drawing lines until they are uniform in width. Make sure the ink does not run under the edge of the T—square or triangle.

Most of the difficulties you will encounter when starting to ink are shown in Fig. 14-7. Many of them can be avoided, or their occurrence greatly reduced, if you keep the pen clean and fill the pen with the quill on the ink bottle cap, Fig. 14-8 (NEVER dip the pen in the ink). Use great care so that you do not accidentally slide your instruments or hand into freshly inked lines.

When inking a tracing, the experienced drafter uses a definite order of working. The order preferred by the author is to ink:
1. All circles, arcs and tangents, starting from the smallest to the largest.
2. Hidden circles and arcs.
3. Irregular curved lines.
4. Straight lines in this order:
   a. Horizontal.

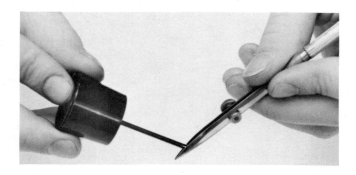

Fig. 14-8. The correct way to fill a ruling pen. NEVER dip the pen in the ink bottle.

b. Vertical.

c. Inclined.

5. Hidden straight lines in the same order listed in 4.

6. All center, extension and dimension lines.

7. Border and section lines.

8. Lettering, dimensions and arrowheads.

TEST YOUR KNOWLEDGE - UNIT 14

(Write answers on a separate sheet of paper.)

1. Why must drawings be duplicated?

2. A blueprint is composed of _____.

3. List the sequence for making blueprints.

4. The diazotype process is often called a _____ print process.

5. How is the diazotype print developed?

6. In microfilming, the original drawing is reduced in size _____.

7. Enlarged prints made from microfilm are called _____.

8. List three basic types of tracing media.

9. When making a tracing, it must be remembered that only a _____, _____ line will print well.

10. When making an ink tracing, the ink used is called _____ ink.

11. _____ is the most usual difficulty encountered when inking.

OUTSIDE ACTIVITIES

1. Make a pencil tracing of one of your drawings and duplicate it by one of the duplicating processes — blueprint, moist print, diazotype print, etc.

2. Make an ink tracing of the same drawing and duplicate it by the same reproduction process. Compare the results.

3. Secure an example of each of the reproduction processes described in this unit. Prepare a bulletin board display around them.

4. Obtain an example of a microfilm aperture card and print made from a microfilm master.

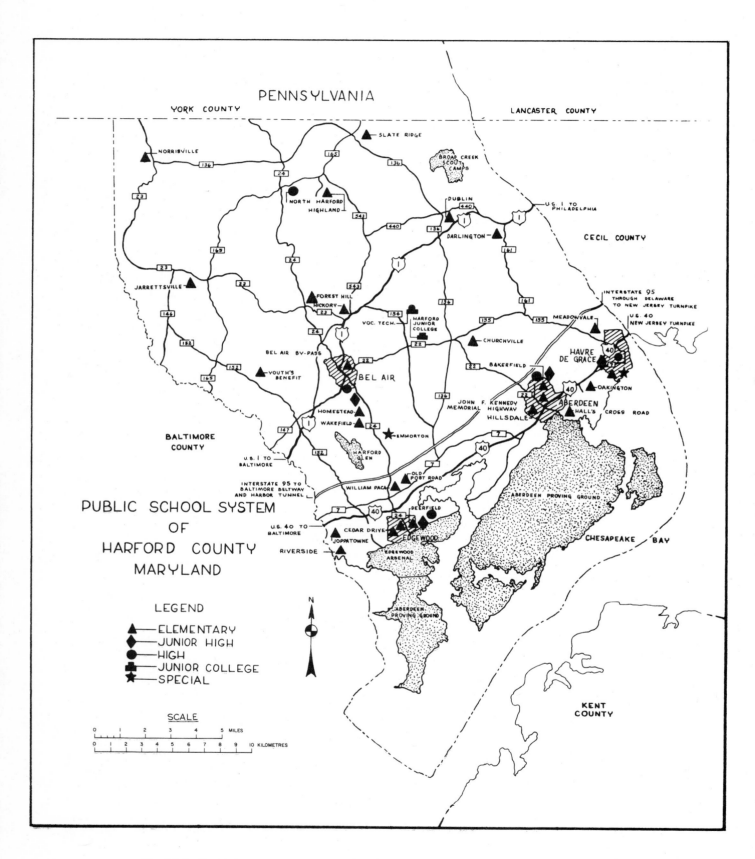

Fig. 15-1. Special purpose map showing the location of schools in Harford County, Maryland (home of the author). Note that both miles and kilometres are used to show the scale of the map.

Unit 15
MAPS

What is a map? We often think of a map as being a graphic picture showing a portion of the earth's surface. However, there are many types of maps. Aeronautical charts or maps provide information for fliers. Airports are shown, as well as compass directions, restricted areas and radio station frequencies needed for navigation. Weather maps let us know what to expect weatherwise. An example of a special purpose map is given in Fig. 15-1.

All of us have used a map of some sort. It may have been a road map to plan a vacation trip or to find the shortest route from one town to another. Or, it may have been in school when studying history or geography.

Professional map makers are called CARTOGRAPHERS. They have the specialized skills required to design and draw maps.

There are times when drafters are called upon to draw maps. This unit will describe the maps they would be expected to prepare.

Aerial photography and the use of space satellites have reduced the time needed to map large land areas. Accuracy also has been improved. However, a great

deal of "legwork" is necessary to map smaller areas. A SURVEYOR, using an instrument called a TRANSIT, Fig. 15-2, establishes the boundaries of the area

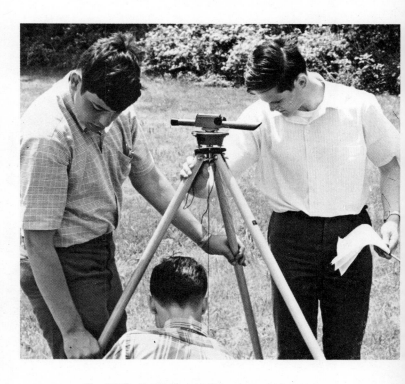

Fig. 15-2. Students setting up a transit.

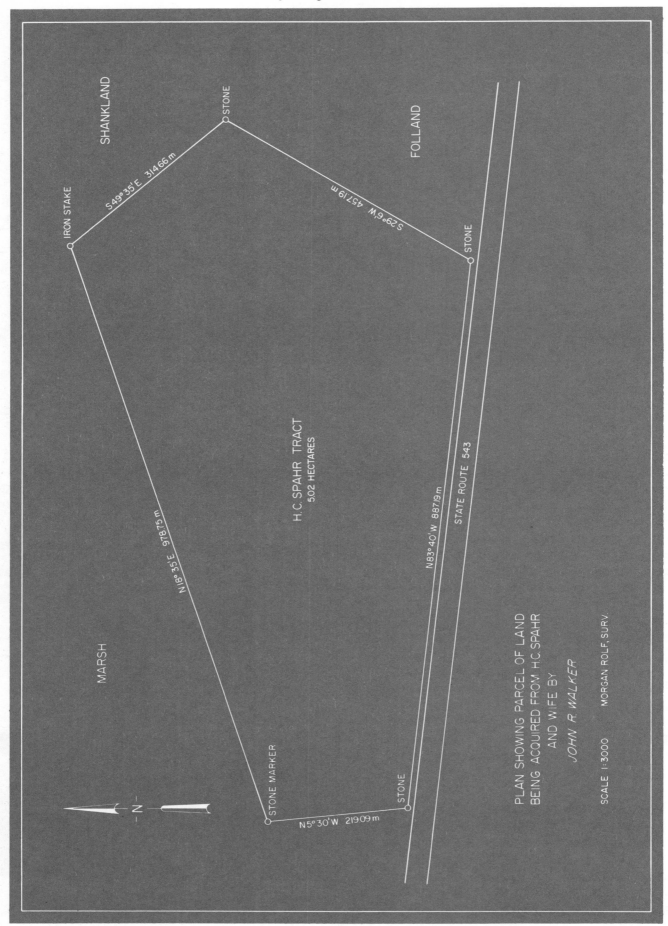

SHANKLAND

STONE

IRON STAKE

S 49° 35' E 314.66 m

S 29° 6' W 457.19 m

FOLLAND

STONE

N 18° 35' E 978.75 m

H.C. SPAHR TRACT

5.02 HECTARES

N 83° 40' W 887.19 m

STATE ROUTE 543

MARSH

STONE MARKER

N 5° 30' W 219.09 m

STONE

N

PLAN SHOWING PARCEL OF LAND
BEING ACQUIRED FROM H.C. SPAHR
AND WIFE BY
JOHN R. WALKER

SCALE 1:3000   MORGAN ROLF, SURV.

Fig. 15-3. Copy of a typical land survey plan. Plan has been metricated.

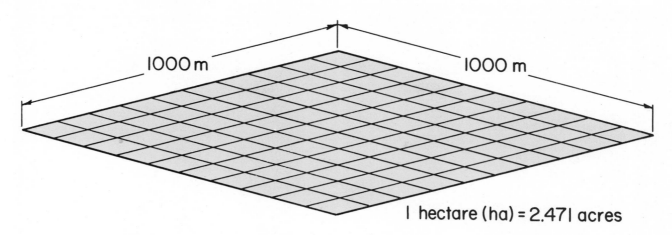

1 hectare (ha) = 2.471 acres

Fig. 15-4. A hectare (ha) is an area that is 100 m long and 100 m wide. It equals about 2.5 acres.

being mapped. Distances are measured with a TAPE and the BEARINGS (angles the lines make with due north and south) are recorded in the form of FIELD NOTES. This information is used to draw maps such as the one shown in Fig. 15-3. North is towards the top of the sheet and is so indicated.

When surveying is metricated, the area of the surveyed land will be given in HECTARES instead of ACRES. See Fig. 15-4. Distances will be noted in METRES and KILOMETRES.

Maps are drawn to scale. The size of the land area being mapped will determine the scale. A drawing of a town lot (called a PLAT) may be drawn to a scale of 1:300. That is, 1.0 mm on the drawing equals 300 mm on the lot. Large land areas may be drawn to a scale of 1:3 500 000 (1.0 cm equals 35.0 km). The U.S. Geological Survey has started mapping the states

in metric. Each map will cover about 4157 square kilometres. They will be printed to a scale of 1:25 000 (4.0 cm equal 1.0 km).

Many maps use a GRAPHIC SCALE, Fig. 15-5, which is easy to interpret. At present, the scale usually is given in kilometres AND miles.

PLAT PLAN

The map of a town lot is called a PLAT or PLOT PLAN, Fig. 15-6. The exact location of the lot and its size is given, as well as how the house or structure is situated on it. The elevation or height of the land above sea level may also be shown as a series of irregular lines (called CONTOUR LINES). These lines are drawn on the map at predetermined differences in elevation. Contour lines may be sketched in lightly, then drawn to the correct line weight.

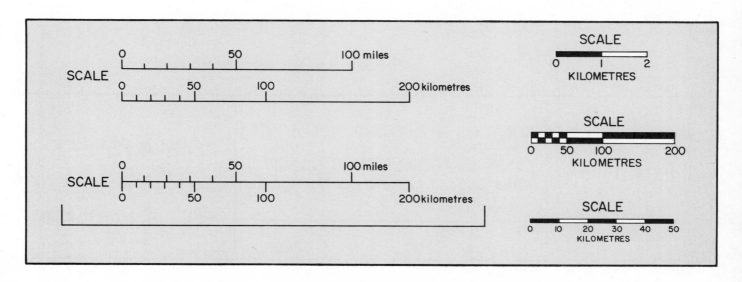

Fig. 15-5. Graphic scales.

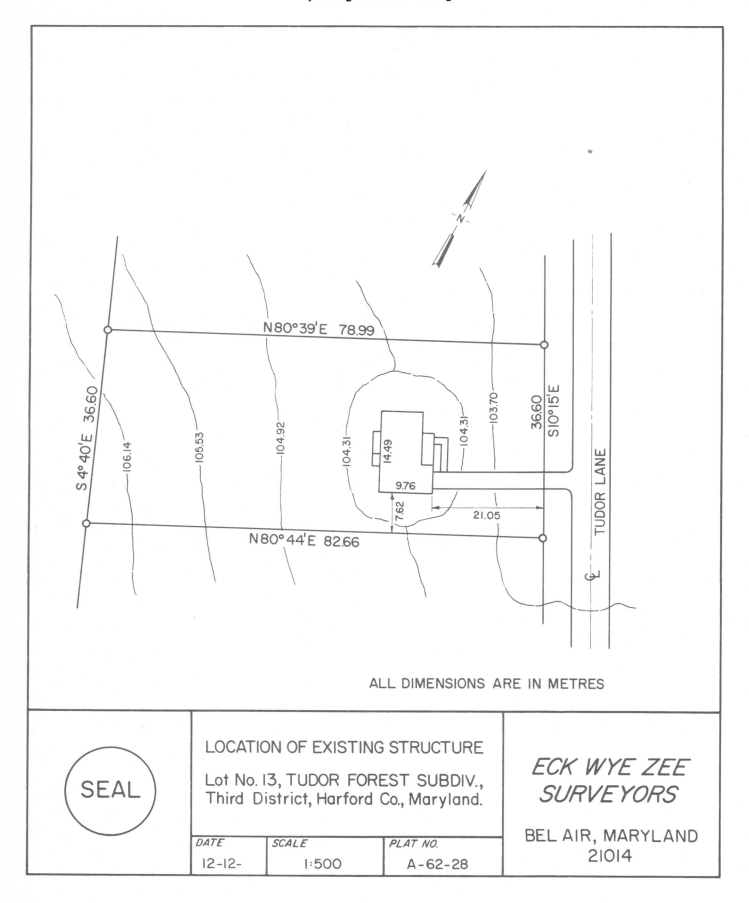

ALL DIMENSIONS ARE IN METRES

LOCATION OF EXISTING STRUCTURE

Lot No. 13, TUDOR FOREST SUBDIV.,
Third District, Harford Co., Maryland.

SEAL

DATE	SCALE	PLAT NO.
12-12-	1:500	A-62-28

*ECK WYE ZEE
SURVEYORS*

BEL AIR, MARYLAND
21014

Fig. 15-6. Plat plan using metric measurements.

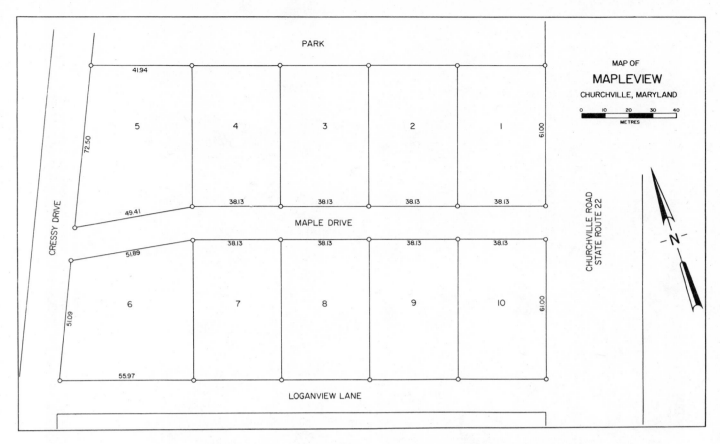

Fig. 15-7. Map of a section of a city showing the layout of streets and lots.

CITY MAP

Street and lot layouts are shown on a CITY MAP, Fig. 15-7. Note that a graphic scale is used on the city map shown to indicate the scale of the map. Also see Fig. 15-5.

TOPOGRAPHICAL MAP

Information about the physical features of land is included on a TOPOGRAPHICAL MAP, Fig. 15-8. Orchards, streams, forests, airports, harbors, roads, bridges, buildings and other land characteristics can be shown. Quite frequently, contour lines are included on a topographic map.

SYMBOLS

By using SYMBOLS, Fig. 15-9, it is possible to include a large amount of information on a map. The symbol is often a stylized drawing that resembles the feature it represents.

TEST YOUR KNOWLEDGE - UNIT 15

(Write answers on a separate sheet of paper.)

1. List some uses you have made of maps.
2. A map is usually thought of as a _____
_____.
3. A _____ uses an instrument called a transit in preparing maps.
4. Why is it impractical to draw a map full size or half size?
5. A plat plan is a map of _____.
6. A city map shows _____.
7. A topographic map shows _____
_____.
8. Topographic features are represented on a map by special symbols. Draw the symbols that represent the following:
   a. Church.   f. Orchards.
   b. House.   g. Corn.
   c. School.   h. Wheat.
   d. Power line.   i. Sand.
   e. Barbed wire fence.

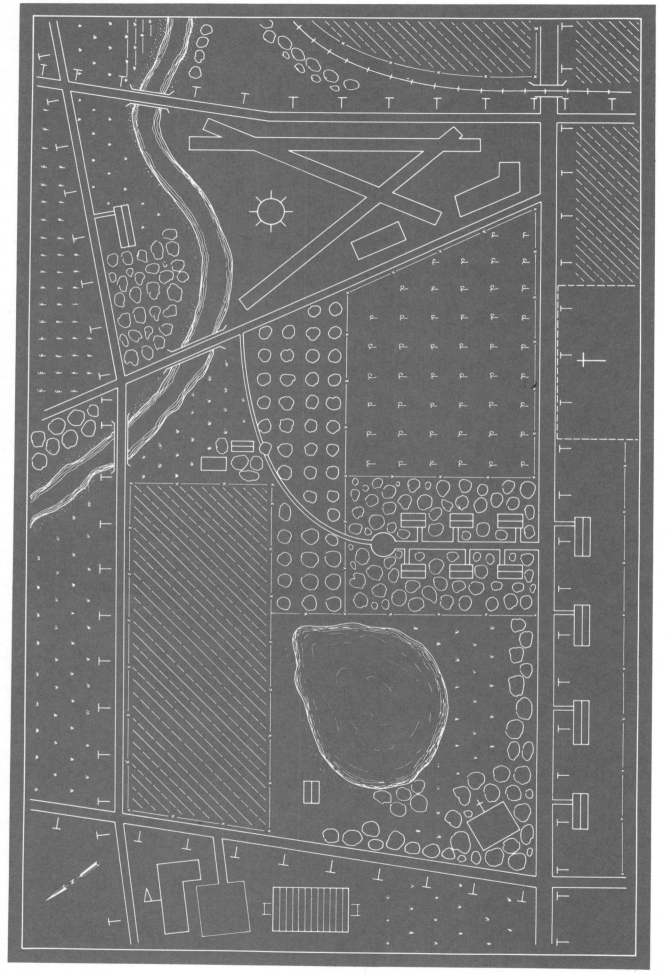

Fig. 15-8. Topographical map.

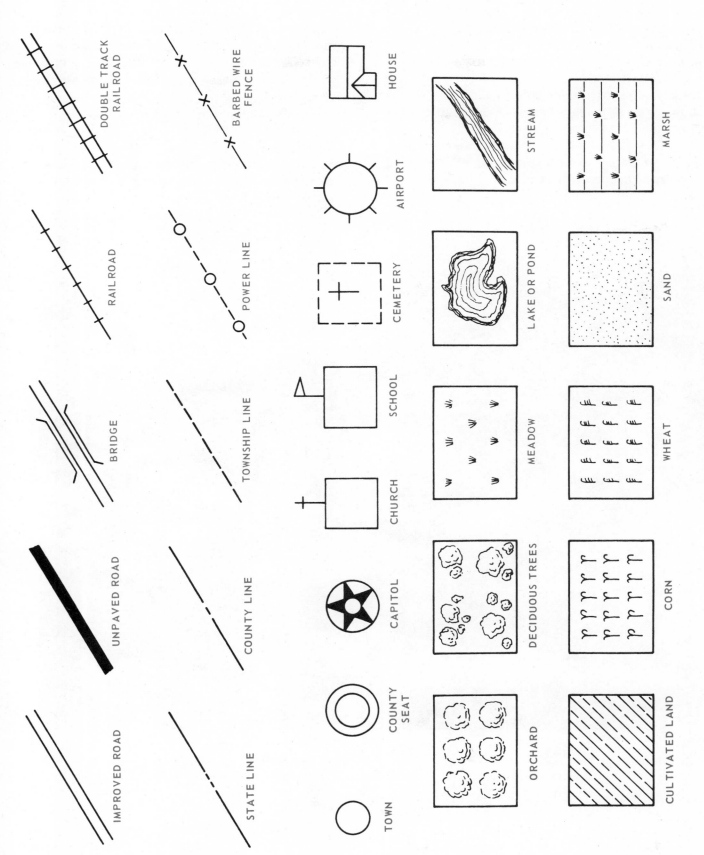

Fig. 15-9. Symbols used on maps.

OUTSIDE ACTIVITIES

1. Secure samples of special purpose maps.
2. Prepare a map showing the school grounds.
3. Make a plat plan of the property on which your home is located.
4. Prepare a map of your neighborhood.
5. Draw a map showing the route you take in coming to school.
6. CLASS PROJECT: Draw a map of the surrounding community and have students draw the routes they travel in coming to school.

7. Secure a road map of your state. Plot the shortest route between your home town and the capitol of your state. If you live in the capitol city, plot the shortest route between it and the next largest city.
8. Prepare a special map that will show the locations of the various schools in your school district.
9. MULTI-GROUP PROJECT: If survey equipment is available, have each group survey the school grounds and compare their results. Also compare their results with an actual survey of the grounds.

Physical features of land such as roads, bridges, forests and streams are included on topographical maps. Often the information is obtained through aerial photography.   (Prestressed Concrete Institute)

Unit 16
CHARTS AND GRAPHS

Industry and education have many uses for charts and graphs. A few of the more widely used kinds are described and shown in this unit.

By using charts and graphs, it is possible to show trends, make comparisons, and measure progress quickly without studying a mass of statistics.

Drafters are often asked to prepare charts and graphs. They take statistical information (figures) and decide upon the most effective way to present this material in an interesting and easily understood manner.

GRAPHS

Kinds of graphs frequently used are:
1. Line Graph.
2. Bar Graph.
3. Circle or Pie Graph (also called Area Graph).

LINE GRAPH

The LINE GRAPH may be used to make comparisons. See Fig. 16-1. Lines that present the information are called CURVES. When only one curve appears on the graph, draw it as a solid line.

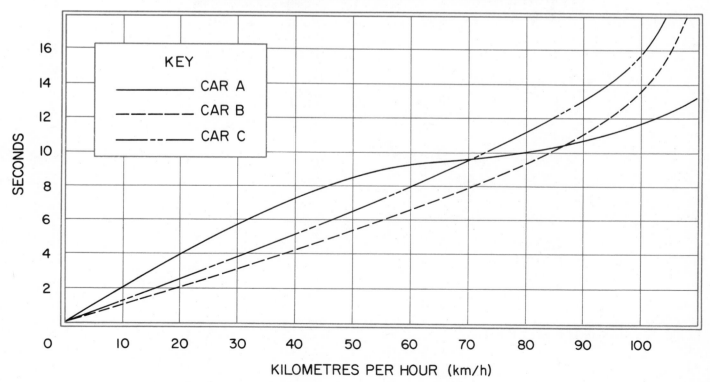

Fig. 16-1. A LINE GRAPH that compares the times needed by three different cars to reach 100 km/h.

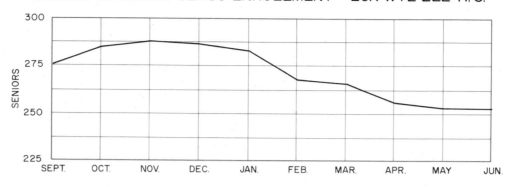

Fig. 16-2. LINE GRAPH showing trends.

When more than one curve is employed, each line should be clearly labeled. A KEY should be included with the graph to show what each curve represents.

A line graph may be used to show trends; that is, what has happened or what may happen. A line graph is illustrated in Fig. 16-2.

BAR GRAPHS

Comparisons between quantities or conditions can also be made using BAR GRAPHS. This particular type of graph is well suited for presenting statistical data. Several different forms of the bar graph are available to the graph maker:

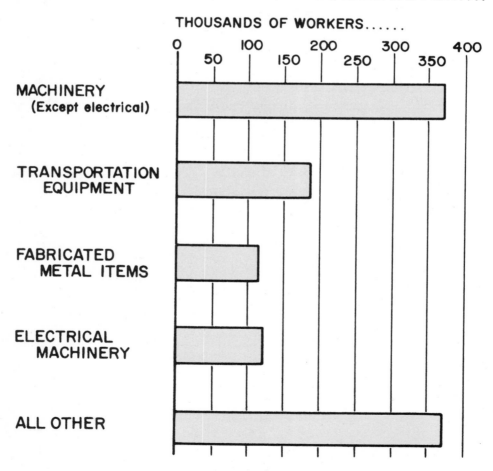

Fig. 16-3. A HORIZONTAL BAR GRAPH that shows where machinists are employed.

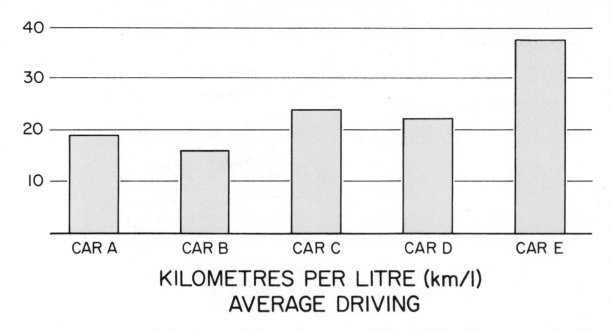

Fig. 16-4. Car economy is shown in this VERTICAL BAR GRAPH.

The HORIZONTAL BAR GRAPH, Fig. 16-3, presents information on a horizontal plane.

With the VERTICAL BAR GRAPH, Fig. 16-4, information is given in a vertical or upright position.

The COMPOSITE BAR GRAPH, Fig. 16-5, can be drawn in either a vertical or horizontal position. This grouping or subdivision of bars compares several items of information in the same graph. It is an effective way of comparing percentages.

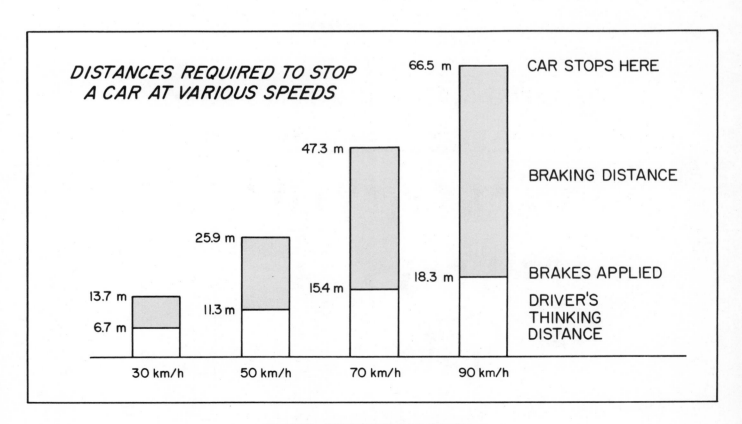

Fig. 16-5. COMPOSITE BAR GRAPH.

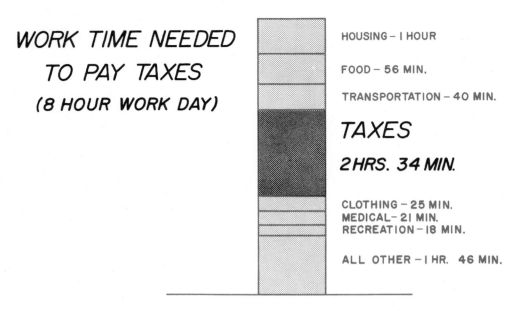

WORK TIME NEEDED
TO PAY TAXES
(8 HOUR WORK DAY)

HOUSING – I HOUR

FOOD – 56 MIN.

TRANSPORTATION – 40 MIN.

TAXES

2 HRS. 34 MIN.

CLOTHING – 25 MIN.
MEDICAL – 21 MIN.
RECREATION – 18 MIN.

ALL OTHER – I HR. 46 MIN.

Fig. 16-6. A 100 PERCENT BAR GRAPH which indicates amount of time one person worked to pay his taxes.

A 100 PERCENT BAR GRAPH, Fig. 16-6, consists of a single rectangular bar. In this type graph, information is presented on a percentage basis.

The PICTORIAL BAR GRAPH, Fig. 16-7, is a variation of the bar graph. It uses pictures to represent the information instead of bars. Pictures tend to make the graph more interesting.

CIRCLE OR PIE GRAPH

The CIRCLE or PIE GRAPH is shown in Fig. 16-8. This is composed of a segmented circle and shows the entire unit divided into comparable parts.

The circle graph is also known as an AREA GRAPH.

TEACHING STAFF – HARFORD CO. SCHOOLS

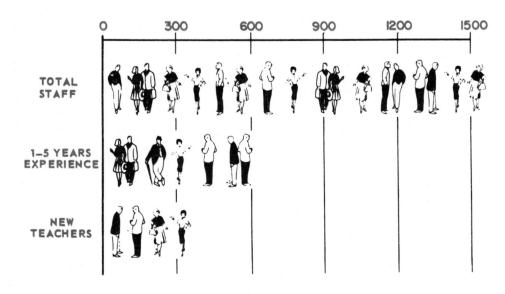

Fig. 16-7. A PICTORIAL BAR GRAPH showing experience of a teaching staff in Harford County, Maryland.

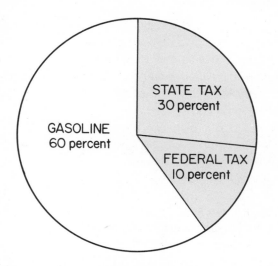

Fig. 16-8. The CIRCLE or PIE GRAPH is sometimes called an AREA GRAPH. This one shows that more than one-third the cost of a gallon of gasoline is state and federal taxes.

CHARTS

CHARTS are another means employed to convey information rapidly. The most familiar of these is the ORGANIZATION CHART, Fig. 16-9. This shows the order of responsibility and the relation of persons and/or positions in an organization.

FLOW CHARTS, Fig. 16-10, may be used to show the sequence, or order of operations, of how a product is manufactured and/or distributed.

The PICTORIAL CHART, Fig. 16-11, is an ideal way of presenting comparisons in an interesting and easily understood manner.

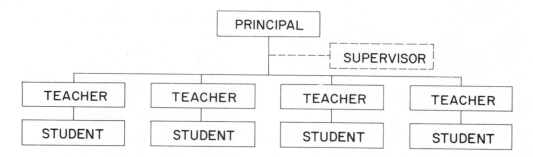

Fig. 16-9. Above. An ORGANIZATION CHART. Fig. 16-10. Below. This FLOW CHART illustrates the sequence a spacecraft used to dock with the lunar module.

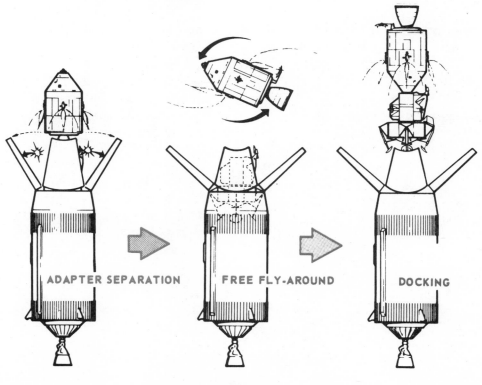

ADAPTER SEPARATION   FREE FLY-AROUND   DOCKING

SUPER JET

DC-3

Fig. 16-11. PICTORIAL CHART comparing the commercial aircraft used in the early 1940's and the commercial jet of today. (Trans World Airlines)

PREPARING CHARTS AND GRAPHS

There are many variations of the graphs presented in this unit. It is up to the ingenuity of the person preparing the material to decide which type will do the best job.

Do not start a graph or chart until ALL of the information has been gathered. Then a rough drawing should be prepared to determine what size presentation will be best.

Use color whenever possible to brighten up your presentation and to make the information stand out.

TEST YOUR KNOWLEDGE - UNIT 16

(Write answers on a separate sheet of paper.)

1. Why are charts and graphs used?
2. List the three most commonly used kinds of graphs.
   a. _____.
   b. _____.
   c. _____.

3. The line that presents the information on a _____ graph is called _____.
4. Of what use is the KEY that should be included when more than one line is used on a graph?
5. Briefly describe each of the following graphs:
   Horizontal Bar Graph.
   Vertical Bar Graph.
   Composite Bar Graph.
   100 Percent Bar Graph.
   Pictorial Bar Graph.
6. The _____ or _____ graph is also known as an Area Graph.
7. Briefly describe each of the following charts:
   Flow Chart.
   Organization Chart.
   Pictorial Chart.
8. Why should color be used on a chart or graph?

OUTSIDE ACTIVITIES

1. Develop a pie graph of how you spend your allowance.
2. Prepare a line graph to show the approximate increase in the price of a particular make of automobile from 1940 to the present time. Use five year steps.
3. Make a 100 percent bar graph showing a breakdown of the cost of a gallon of gasoline, including cost of the gasoline and state, federal and local taxes.
4. Draw a pictorial bar graph showing how far an automobile will travel after the brakes are applied at 25 mph, 45 mph, 60 mph and 75 mph.
5. Make an organization chart of the pupil personnel system used in your drafting room or Industrial Arts shops.
6. Develop a picture chart showing how the size of a particular make of automobile has increased, then decreased in the past 20 years.
7. Make a picture graph showing the enrollment in each grade of your school. Let each symbol represent 25 students.
8. Design a flow chart showing how a simple stool could be manufactured in the Industrial Arts shop.
9. Make a line graph showing the enrollment of the Industrial Arts classes in your school.

Unit 17
MANUFACTURING
PROCESSES

Metals and other materials are available in a variety of shapes and sizes. See Fig. 17-1. Since the purpose of most drawings is to describe a part or product that is to be manufactured, it is important that the drafter have an understanding of how the materials can be cut, shaped, formed and fabricated. He should also know which material is best suited for a particular application or product.

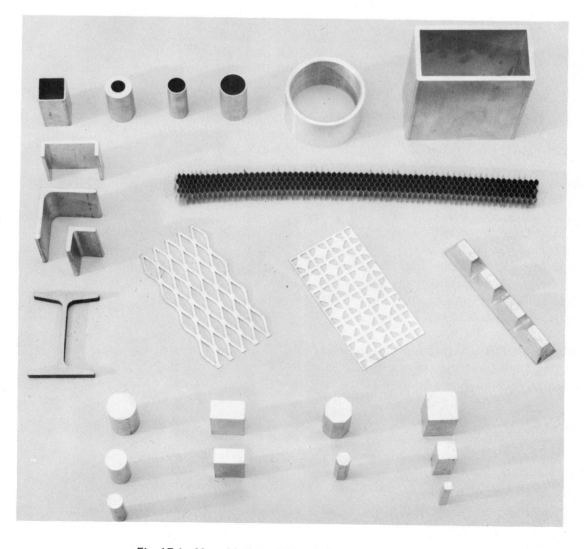

Fig. 17-1. Many kinds and shapes of metal are available.

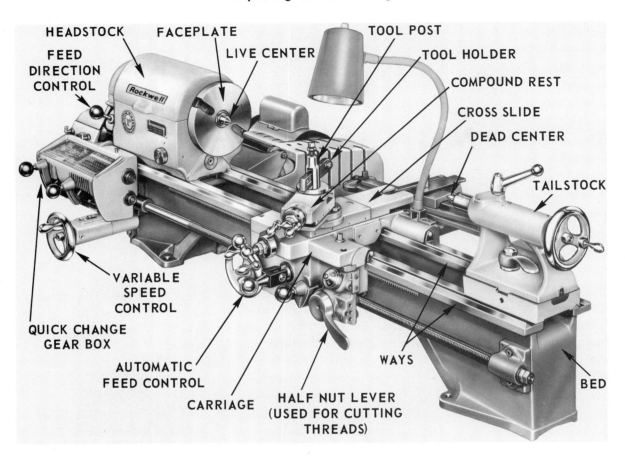

Fig. 17-2. Metal cutting lathe. (Rockwell Mfg. Co.)

MACHINE TOOLS

"A machine tool is a power driven machine, not portable by hand, used to shape or form metal by cutting, impact, pressure, electrical techniques, or by a combination of these processes."[1]

The world of today could not exist without MACHINE TOOLS. They produce the accurate and uniform parts needed for the assembly line and make mass production techniques possible.

Machine tools are manufactured in a large range of styles and sizes. Only basic tools and manufacturing techniques will be covered in this unit.

LATHE

The LATHE, Fig. 17-2, is one of the oldest and most important of all machine tools. It operates on the principle of the work being rotated against the

edge of the cutting tool, Fig. 17-3. The cutting tool can be controlled and can be moved lengthwise and across the face of the material being machined (turned).

Operations other than turning can be performed on the lathe. It is possible to drill, bore, Fig. 17-4, ream and cut threads and tapers.

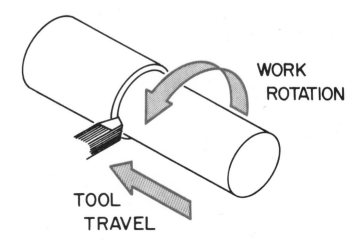

Fig. 17-3. The operating principle of the lathe.

[1] National Machine Tool Builders' Association
7901 West Park Drive
McLean, Virginia 22102

196

Fig. 17-4. This boring operation by lathe involves machining an internal surface. (Clausing)

Fig. 17-5. Turret lathe performs drilling operation. (Clausing)

There are many variations of the basic lathe. The TURRET LATHE, Fig. 17-5, is used when a number of identical parts must be machined. It is a conventional lathe fitted with a six-sided tool holder called a TURRET. Different cutting tools fitted in the turret rotate into position for the machining operations.

Other lathes range in size from the small lathe used by the instrument makers and watchmakers, Fig. 17-6, to the large lathes that machine the forming rolls for steel mills. See Fig. 17-7.

DRILL PRESS

The DRILL PRESS, Fig. 17-8, probably is the best known of the machine tools. On it, a cutting tool called a TWIST DRILL is rotated against the work with sufficient pressure to cut its way through the material, Fig. 17-9.

Other operations which can be performed on the drill press include: REAMING, Fig. 17-10, finishing a drilled hole to close tolerances; COUNTERSINKING, cutting a chamfer on a hole so a flat head fastener can be inserted; TAPPING, the process of cutting internal threads.

MILLING MACHINE

The MILLING MACHINE, Fig. 17-11, is quite a versatile machine. It can be used to machine flat and irregularly shaped surfaces, drill, bore and cut gears.

The VERTICAL MILLING MACHINE, Fig. 17-12, differs from the HORIZONTAL MILLING MA-

Fig. 17-6. Partial assembly of a plate and gear train of a wristwatch being held by tweezers. The gear shafts and pins were made on a lathe. (Bulova Watch)

Fig. 17-7. Lathe used to machine forming rolls for a steel mill.

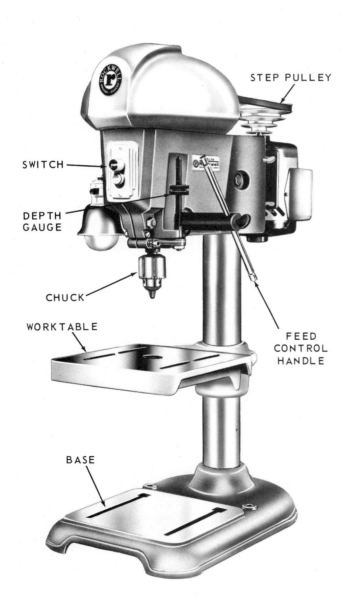

Fig. 17-8. A bench model drill press.

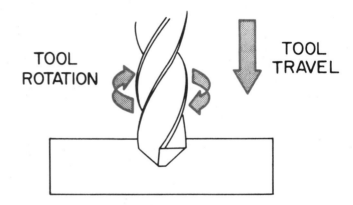

Fig. 17-9. The operating principle of the drill press.

CHINE shown in Fig. 17-11 in that the cutter is mounted vertical to the worktable. Metal is removed by means of a rotating cutter that is fed into the moving work. See Fig. 17-13.

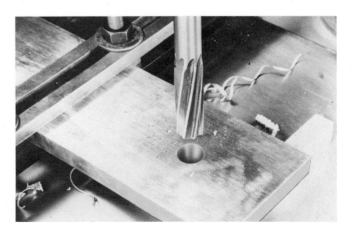

Fig. 17-10. Reaming produces a more accurate hole than drilling. First, the hole must be drilled slightly smaller than the required finished hole size.

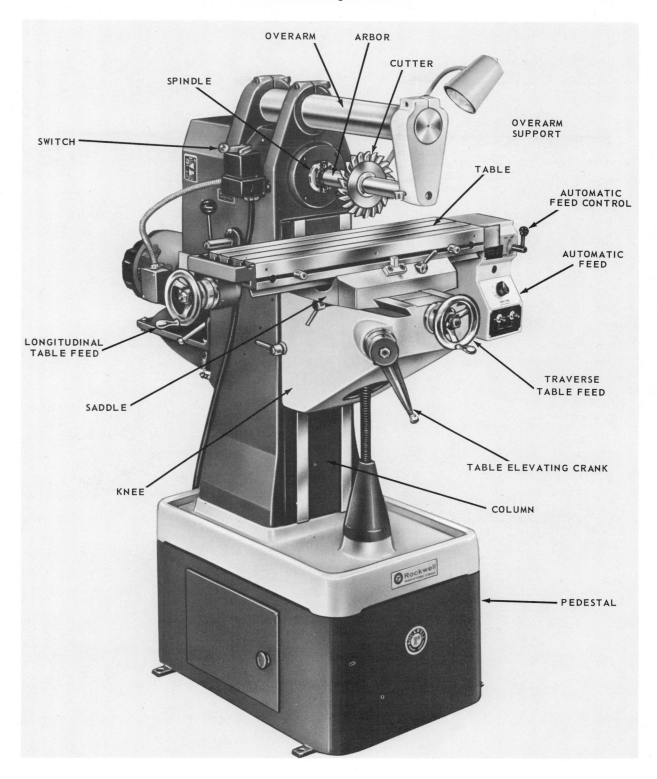

SPINDLE

SWITCH

OVERARM

ARBOR

CUTTER

OVERARM
SUPPORT

TABLE

AUTOMATIC
FEED CONTROL

AUTOMATIC
FEED

LONGITUDINAL
TABLE FEED

SADDLE

KNEE

TRAVERSE
TABLE FEED

TABLE ELEVATING CRANK

COLUMN

PEDESTAL

Fig. 17-11. Horizontal milling machine.   (Rockwell Mfg. Co.)

Other types of milling machines are available. They vary in size and by types of controls. The controls may be activated mechanically, electrically, hydraulically or by a combination of all three.

SHAPER, PLANER AND BROACH

Where the lathe is used to shape metal to a circular shape, planing machine tools — the SHAPER, PLAN-

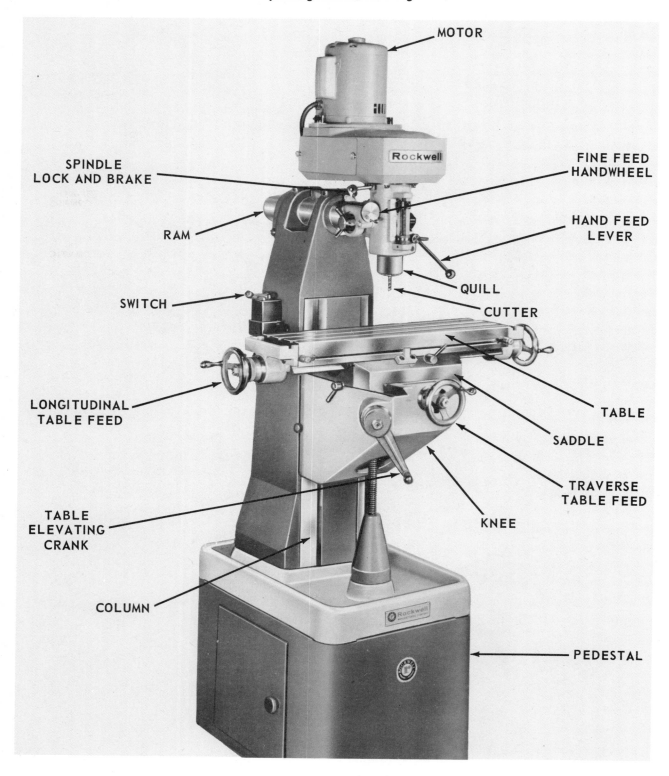

MOTOR

Rockwell

SPINDLE
LOCK AND BRAKE

FINE FEED
HANDWHEEL

RAM

HAND FEED
LEVER

QUILL

SWITCH

CUTTER

LONGITUDINAL
TABLE FEED

TABLE

SADDLE

TRAVERSE
TABLE FEED

TABLE
ELEVATING
CRANK

KNEE

COLUMN

PEDESTAL

Fig. 17-12. The vertical milling machine. (Rockwell Mfg. Co.)

ER and BROACH — are employed to machine flat surfaces.

The SHAPER, Fig. 17-14, is considered too slow for most mass production operations. Much of the work formerly done on the shaper is being done faster and more economically on the milling machine and broaching machine.

A shaper works as shown in Fig. 17-15. A planer differs in that the cutter remains stationary while the work travels back and forth. See Fig. 17-16.

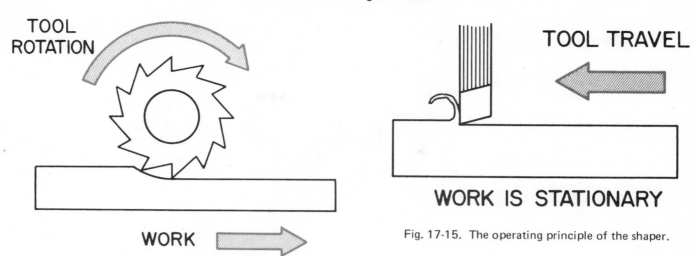

Fig. 17-13. The operating principle of the milling machine.

Fig. 17-15. The operating principle of the shaper.

BROACHING, Fig. 17-17, is similar to shaping, but instead of a single cutting tool advancing after each stroke across the work, the broach is a long tool with many cutting teeth. Each tooth has a cutting edge a few thousandths of an inch higher than the one before and increases in size to the exact finished size required. The broach is pushed or pulled across the work.

The broach is ideal for producing keyways and irregular shaped openings.

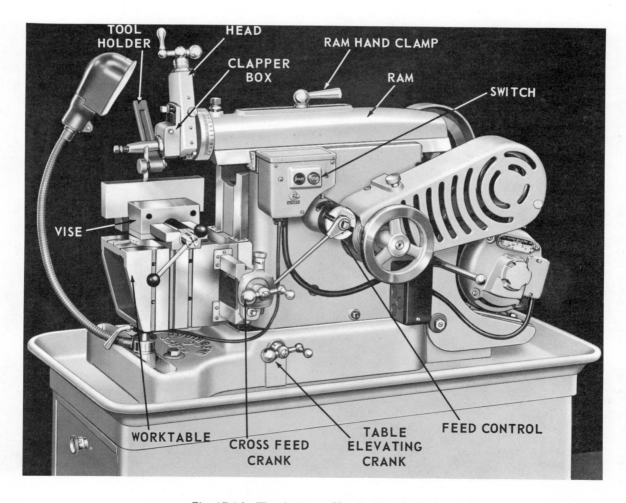

Fig. 17-14. The shaper. (South Bend Lathe)

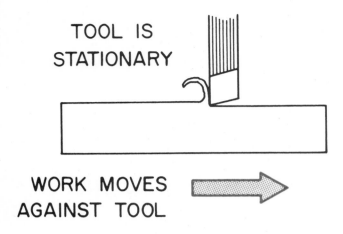

TOOL IS
STATIONARY

WORK MOVES
AGAINST TOOL

Fig. 17-16. The operating principle of the planer.

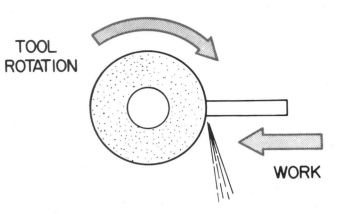

TOOL
ROTATION

WORK

Fig. 17-18. The operating principle of the grinder.

GRINDING

GRINDING, Fig. 17-18, is an operation that removes material by rotating an abrasive wheel against the work. The PEDESTAL GRINDER, Fig. 17-19, is the simplest and most widely used grinding machine.

A grinding machine like the SURFACE GRINDER, Fig. 17-20, is used to put fine finishes on flat surfaces and to finish them to close tolerances.

With the CYLINDRICAL GRINDER, Fig. 17-21, it is economically feasible to machine hardened steel parts to tolerances of 0.000 25 mm (1/100,000 in.) with extremely fine surface finishes.

SHEARING AND FORMING TECHNIQUES

SHEARING is a process where the material (usually in sheet form) is cut to shape, using action similar to cutting paper with scissors.

STAMPING is divided into two separate classifications: CUTTING and FORMING. The cutting opera-

tion is also known as BLANKING, Fig. 17-22. It involves cutting flat sheets to the shape of the finished part. FORMING is a process where flat metal blanks are given three-dimensional form. See Fig. 17-23.

With PRESSING, metal is shaped by using pressure and sometimes heat to form it to the required shape.

CASTING TECHNIQUES

Materials can also be given shape and form by reducing it to a molten state and pouring it into a mold of the desired shape. This is called casting.

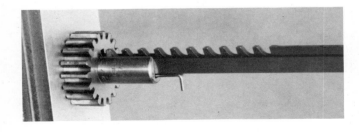

Fig. 17-17. A broach being used to cut a keyway in a gear. Note the number of teeth on the tool.

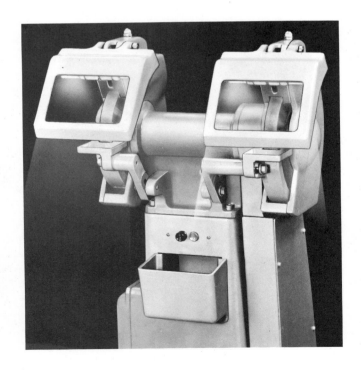

Fig. 17-19. Pedestal grinder. (South Bend Lathe)

202

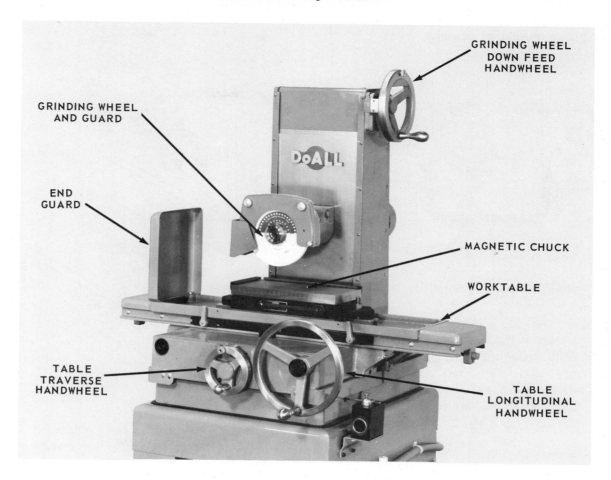

Fig. 17-20. Manually operated surface grinder. (The DoALL Co.)

Fig. 17-21. Closeup of a cylindrical grinder, showing part being ground and abrasive wheel. (Landis Tool Co.)

SAND CASTING

In the SAND CASTING PROCESS, Fig. 17-24, the mold that gives the molten metal shape is made of sand. It is one of the oldest metal forming techniques known.

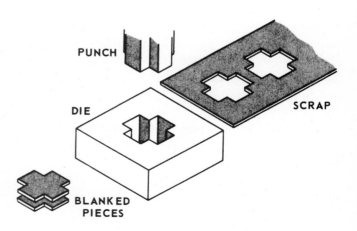

Fig. 17-22. Blanking operation. Cutting flat sheets.

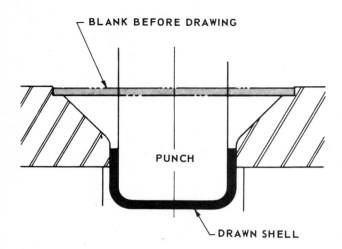

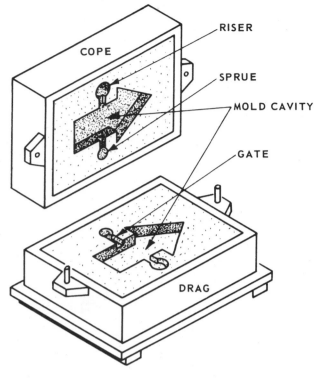

Fig. 17-23. Forming is a process where flat metal blanks are given three-dimensional form.

The sand mold is made by packing sand in a box called a FLASK, around a PATTERN of the shape to be cast. When the pattern has been DRAWN (removed) from the mold and the mold halves, called the COPE and the DRAG, are reassembled, a cavity remains.

Before assembling the mold, openings called SPRUES, RISERS and GATES are made in the mold. The sprue is the opening into which the molten metal is poured. The risers allow the hot gases to escape. The gates are trenches that run from the sprues and risers to the mold cavity and permit the molten metal to reach the cavity.

Because metal contracts as it cools, PATTERNS, Fig. 17-25, must be made oversize to allow for shrinkage.

PERMANENT MOLD CASTING

Some molds used for metal casting are made from metal. These molds are called PERMANENT MOLDS because they do not have to be destroyed to remove the casting. The process produces castings with a fine surface finish and high accuracy, Fig. 17-26.

Fishing sinker and toy soldier molds are familiar examples of products made using the permanent mold process.

DIE CASTINGS

DIE CASTING, Fig. 17-27, is a process where molten metal is forced into a DIE or MOLD under

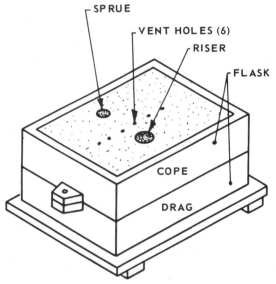

Fig. 17-24. Parts of a sand mold.

pressure. The pressure is maintained until the metal solidifies and the casting is removed, Fig. 17-28. The mold or die is made of metal.

After the casting has been removed, the die is closed and the cycle is repeated. To speed up this cycle, a die casting machine is used. Die cast parts (die castings) have good dimensional accuracy. When properly designed, a die casting requires little finish machining.

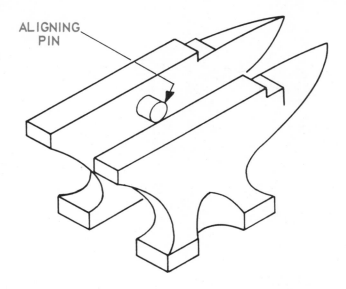

ALIGNING PIN

Fig. 17-25. Typical two-piece pattern used to make cavity in mold.

Fig. 17-26. This piston was cast in a permanent mold in the same way you would cast toy soldiers and fishing sinkers.

Fig. 17-27. Small gas engine. The crankcase, cylinder head and crankcase cover were die cast.

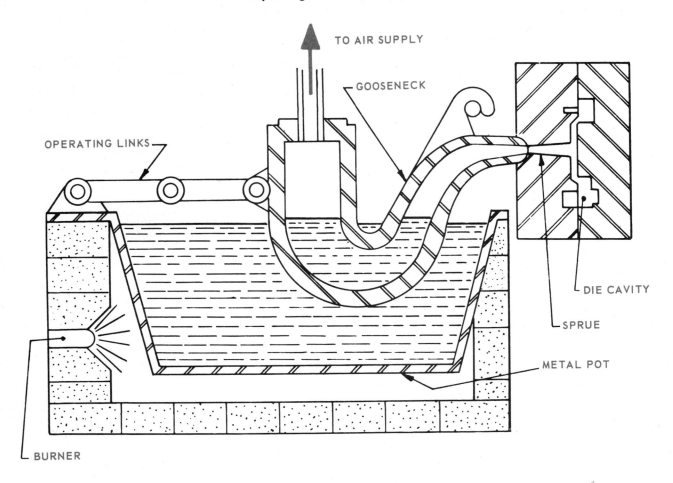

TO AIR SUPPLY

GOOSENECK

OPERATING LINKS

DIE CAVITY

SPRUE

METAL POT

BURNER

Fig. 17-28. Diagram of a die casting machine. The metal is forced into the die cavity (mold) under pressure to make a casting that is denser than a sand casting.

TEST YOUR KNOWLEDGE - UNIT 17

(Write answers on a separate sheet of paper.)

1. Make a sketch showing how a lathe operates.
2. Make a sketch showing how a drill press operates.
3. Make a sketch showing how a milling machine operates.
4. What is a machine tool?
5. What are some other operations that can be performed on a drill press?
   a. _____.   b. _____.
6. List four machines used to machine flat surfaces.
   a. _____.   c. _____.
   b. _____.   d. _____.
7. Make a sketch showing how the surface grinder operates.
8. A _____ is used to make the mold cavity in a sand mold.
9. It is made slightly larger than the finished casting because metal _____.

10. How do sand castings differ from castings made in permanent molds and die castings?
11. Make a sketch showing a typical sand mold.

OUTSIDE ACTIVITIES

1. Prepare a bulletin board using clippings which illustrate basic machine tools.
2. Secure samples of work made on a lathe.
3. Secure samples of work machined on a drill press.
4. Secure samples of work machined on a milling machine.
5. Secure samples of work machined on a surface grinder.
6. Secure samples of work cast in a sand mold.
7. Secure samples of work made by die casting.
8. Bring into class an example of a permanent mold such as a sinker mold or toy soldier mold. Discuss how the item is cast.

Unit 18
WELDING DRAWINGS

WELDING is a widely used industrial technique for fabricating metal pieces. It is a method of joining by heating metals to a high temperature which causes them to melt and fuse together. The high temperatures are generated by electricity (arc welding) or by burning gases (usually oxyacetylene). See Fig. 18-1.

Fig. 18-1. In the manufacturing, construction and service industries, welding is a widely used metal fabricating technique.   (Union Carbide Corp., Linde Div.)

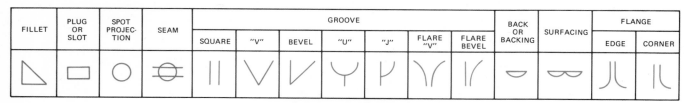

FILLET	PLUG OR SLOT	SPOT PROJEC-TION	SEAM	GROOVE							BACK OR BACKING	SURFACING	FLANGE	
				SQUARE	"V"	BEVEL	"U"	"J"	FLARE "V"	FLARE BEVEL			EDGE	CORNER

BASIC ARC AND GAS WELD SYMBOLS

WELD ALL AROUND	FIELD WELD	MELT-THRU	CONTOUR		
			FLUSH	CONVEX	CONCAVE

SUPPLEMENTARY WELD SYMBOLS

Fig. 18-2. Welding symbols.

The American Welding Society (AWS) has developed a series of symbols which tell the welder exactly what to do.

The symbols are included on drawings where the assemblies require welding. Basic arc and gas welding symbols are shown in Fig. 18-2.

USING WELDING SYMBOLS

Before welding symbols can be used effectively, the drafter must be familiar with the various types of joints used in welding, Fig. 18-3.

The location of the arrow with respect to the joint is very important when using the welding symbol to specify the required weld. The side of the joint indicated by the arrow is considered the ARROW SIDE. The side opposite the arrow side of the joint is considered the OTHER SIDE of the joint.

When the weld is to be made on the ARROW SIDE of the joint, the weld symbol is placed on the reference line so it is TOWARD THE READER, Fig. 18-4. A weld on the OTHER SIDE of the joint is specified by placing the weld symbol on the reference line AWAY FROM THE READER, Fig. 18-5. Welds on both sides of the joint are specified by placing the

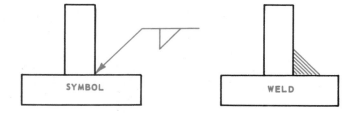

Fig. 18-4. Weld symbol that indicates that the weld is to be made on the ARROW SIDE of the joint. Note the weld symbol is TOWARD the reader.

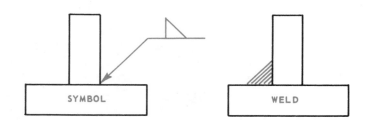

Fig. 18-5. Weld symbol indicates that the weld is to be made on the OTHER SIDE of the joint.

symbol on BOTH SIDES of the reference line. Note positioning of symbol in Fig. 18-6.

A weld that is made all of the way around the joint is specified by drawing a circle at the point where the arrow is bent, Fig. 18-7.

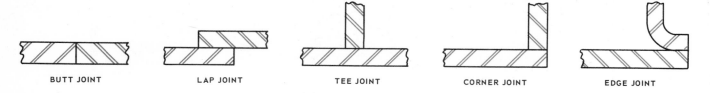

BUTT JOINT   LAP JOINT   TEE JOINT   CORNER JOINT   EDGE JOINT

Fig. 18-3. Basic joints used in welding.

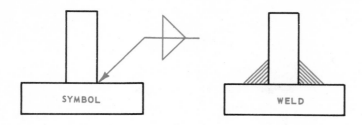

Fig. 18-6. Weld symbol indicates that the weld is to be made on BOTH SIDES of the joint.

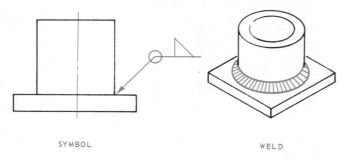

Fig. 18-7. Weld symbol that indicates a weld to be made ALL AROUND the joint.

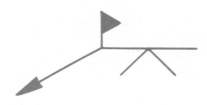

Fig. 18-8. Field welding symbol for "V" groove weld.

FIELD WELDS are specified as shown in Fig. 18-8. The basic weld symbol where the arrow is bent replaces the solid circle used in the past. Field welds are welds made when the parts are assembled away from the shop.

Weld size is placed to the left of the symbol, Fig. 18-9. The figure that indicates the length of the weld is placed to the right of the symbol.

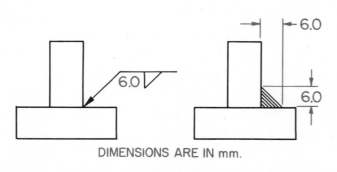

DIMENSIONS ARE IN mm.

Fig. 18-9. How weld size is indicated.

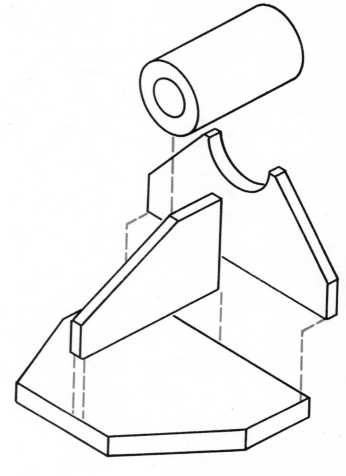

Fig. 18-10. A welded assembly is composed of several different pieces.

DRAWINGS FOR ASSEMBLIES TO BE FABRICATED BY WELDING

Items that are to be assembled by welding usually are composed of several different pieces, Fig. 18-10. The individual pieces are seldom cut to size, welded, and machined in the same general area. Therefore, several drawings usually are required; each of which provides information on a specific operation.

For simple jobs, all of the needed information generally can be included on a single drawing, Fig. 18-11. Note the use of typical welding symbols.

In review, remember that the welding symbols placed on a drawing provide a means for giving complete and specific welding information on the part to be welded. The American Welding Society developed and standardized the basic symbols shown in Fig. 18-2. For complete coverage, the latest copy of Standard Welding Symbols should be obtained from AWS.

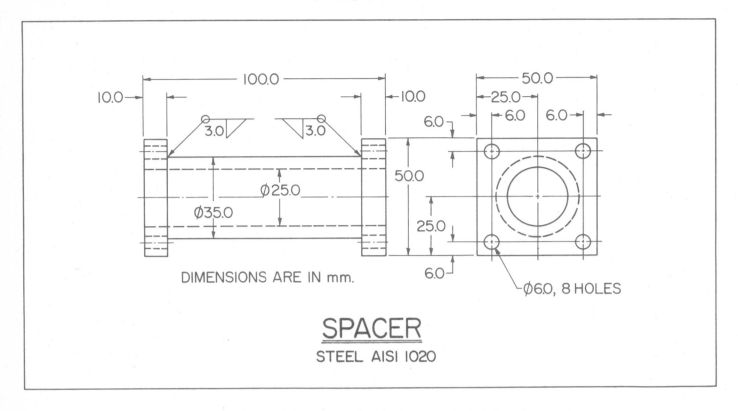

DIMENSIONS ARE IN mm.

SPACER
STEEL AISI 1020

Fig. 18-11. Cutting, welding and machining details are included on this drawing.

TEST YOUR KNOWLEDGE - UNIT 18

(Write answers on a separate sheet of paper.)

1. How are the high temperatures needed for welding generated?
2. Welding symbols were developed to:
   a. Lessen the possibilities of the wrong type of weld being made.
   b. Tell the welder exactly what to do.
   c. Eliminate "hit or miss" welds.
   d. All of the above.
3. Sketch the symbol for a fillet weld on the arrow side of the joint.
4. Sketch the symbol for a fillet weld on the other side of the joint.
5. Sketch the symbol for a fillet weld on both sides of the joint.
6. Sketch the symbol for a fillet weld all of the way around the joint.
7. What is a field weld?
8. Why are several different drawings needed for a job that is made up of several pieces and assembled by welding?

OUTSIDE ACTIVITIES

1. Secure samples of welded joints and mount them on a display board. Label the samples and add the proper drafting symbols to the display.
2. Visit a local industry that makes extensive use of welding and get samples of the work they produce.
3. Invite a professional welder to the school shop to demonstrate the safe and proper way to gas weld and to electric arc weld.

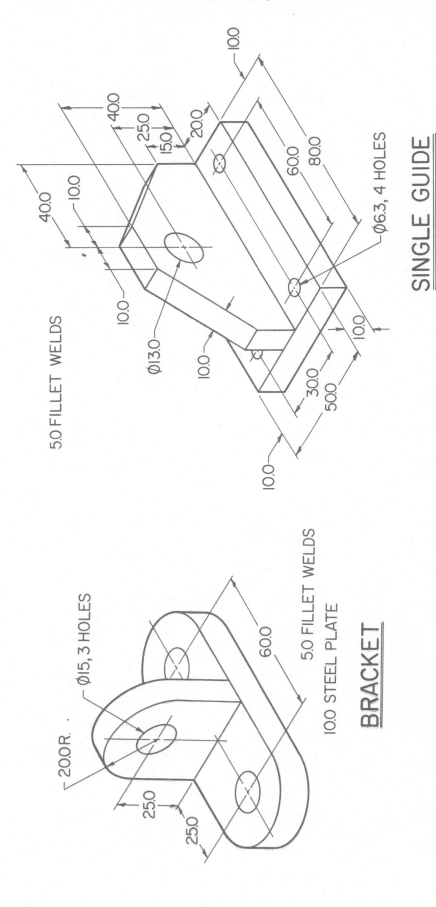

SINGLE GUIDE

5.0 FILLET WELDS

Ø13.0

Ø6.3, 4 HOLES

BRACKET

Ø15, 3 HOLES

20.0R.

5.0 FILLET WELDS

10.0 STEEL PLATE

PROBLEM 18–1. Left. BRACKET. Prepare a drawing showing the necessary views to fabricate it. Use the appropriate welding symbol. PROBLEM 18–2. Right. SINGLE GUIDE. Prepare a drawing with the information necessary to fabricate it by welding.

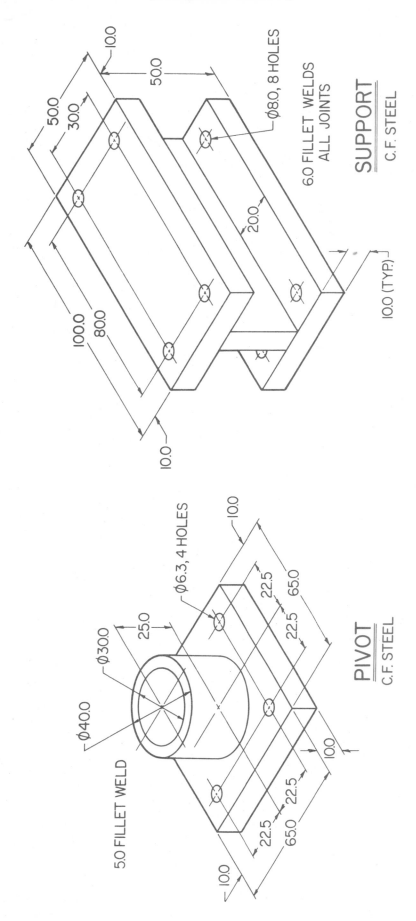

PROBLEM 18–3. Left. PIVOT. The fillet weld is to be made all around the joint. PROBLEM 18–4. Right. SUPPORT. Prepare a drawing showing the views necessary to fabricate the SUPPORT.

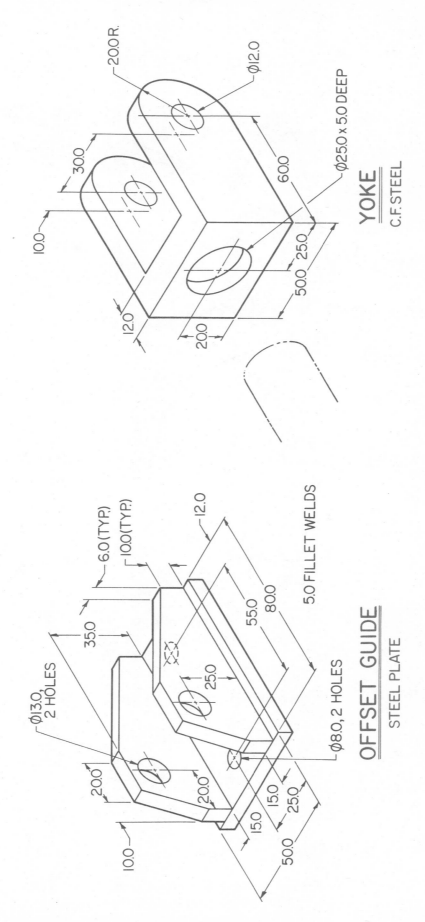

YOKE
C.F. STEEL

OFFSET GUIDE
STEEL PLATE

PROBLEM 18—5. Left. OFFSET GUIDE. The weld is to be made on both sides of each vertical piece. PROBLEM 18—6. Right. YOKE. This part is to be welded to a 25.0 mm diameter shaft at the job site.

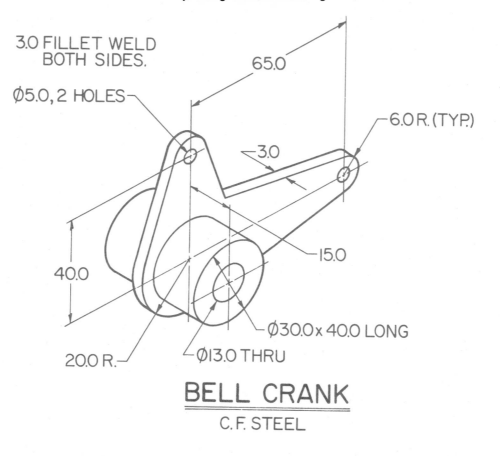

3.0 FILLET WELD
BOTH SIDES.

Ø5.0, 2 HOLES

65.0

6.0 R. (TYP.)

3.0

15.0

40.0

Ø30.0 x 40.0 LONG

20.0 R.

Ø13.0 THRU

BELL CRANK
C.F. STEEL

PROBLEM 18–7. BELL CRANK. A 3.0 mm fillet weld is specified around both sides of the joint.

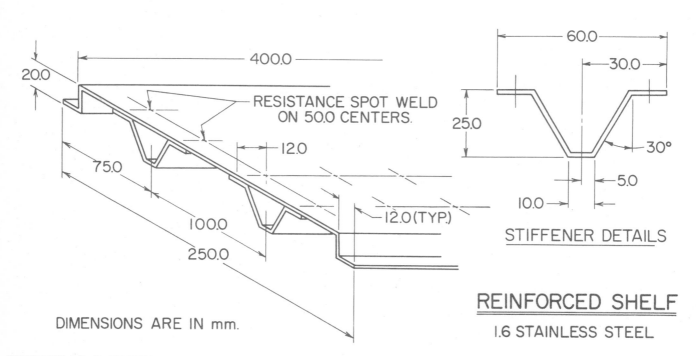

20.0

400.0

RESISTANCE SPOT WELD
ON 50.0 CENTERS.

12.0

75.0

12.0 (TYP.)

100.0

250.0

DIMENSIONS ARE IN mm.

60.0

30.0

25.0

30°

5.0

10.0

STIFFENER DETAILS

REINFORCED SHELF
1.6 STAINLESS STEEL

PROBLEM 18–8. REINFORCED SHELF. Stiffeners are attached to the shelf with resistance spot welds. Welds are placed on 50.0 mm centers on each flange of the stiffener.

Unit 19
FASTENERS

A fastener is a device used to hold two or more objects or parts together. Fasteners include screws, rivets, nails, nuts and bolts, etc., used extensively to assemble manufactured items, Fig. 19-1.

THREADED FASTENERS

Threads have many applications. They are employed to:
1. Make adjustments.
2. Transmit motion.
3. Assemble parts.
4. Apply pressure.
5. Make measurements.
How many examples of these applications can you name?

Fasteners that fall into the threaded category use the wedging action of the thread to hold things together.

Engineers, drafting personnel and mechanics must now become familiar with metric based threads.

Metric fasteners are made in coarse and fine thread series, Fig. 19-2, much like the American Standard for Unified Screw Threads coarse (UNC) and fine UNF) series. However, metric fine threads are used primarily in precision instruments and are not commonly available.

Fig. 19-2. Metric fasteners are made in both fine and coarse thread series. The bolts shown are the same diameter and length. How do they differ?

Fig. 19-1. These VTOL (Vertical Take Off and Landing) jets are being manufactured for the U.S. Marines. Metric standards are used. More than one million fasteners are used in each aircraft. (Hawker Siddeley Aviation Limited)

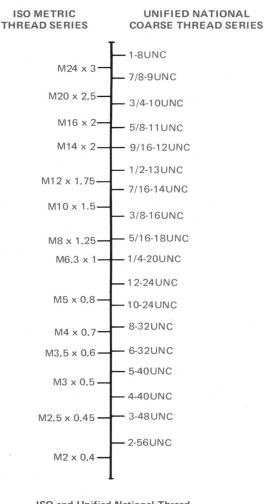

ISO and Unified National Thread
Series ARE NOT INTERCHANGABLE

Fig. 19-3. A comparison of ISO metric coarse pitch and Unified Coarse (UNC) inch based thread sizes. Even though several of them seem to be the same size, THEY ARE NOT INTERCHANGEABLE.

IDENTIFYING METRIC BASED FASTENERS

Industry anticipates that problems could occur during the metrication of threaded fasteners. At present, more than two million different kinds, shapes and sizes of inch based threaded fasteners are made. Can you imagine what will happen when metric based threaded fasteners are added to this large list? Both inch based and metric based fasteners will have to be kept in inventory for many, many years. Both inch based and metric based wrenches will be needed for the same period of time.

A comparison of ISO metric coarse pitch and Unified coarse (UNC) inch based thread sizes is shown in Fig. 19-3.

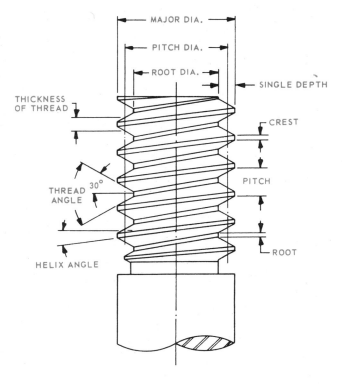

Fig. 19-4. Metric and inch based threads have exactly the same shape. Shown are the parts of a screw thread.

Note in Fig. 19-4 that metric based and inch based threads have the same basic profile (shape). While they appear to be similar in diameter and pitch, THEY ARE NOT INTERCHANGEABLE. That is, a metric bolt cannot be used with an inch based nut that appears to have the same thread size as the bolt.

The user may have trouble telling a metric bolt from a similar inch based fastener. Some simple way will have to be devised to identify the metric fastener.

An attempt was made to color code metric fasteners by a dying process. This proved too costly. Note, too, that the larger hex headed metric fasteners often have the thread diameter imprinted on the head. However, this is impractical for small size fasteners.

At present in the United States, the imprinted hex head and a unique 12-element spline head are being considered for use. See Fig. 19-5.

Also under consideration is retention of the Phillips recessed head for inch based drive fasteners. The POZIDRIV.® recessed head, Fig. 19-6, will be used solely on metric fasteners. The POZIDRIV® resembles the Phillips configuration, but the two should not be used interchangeably.

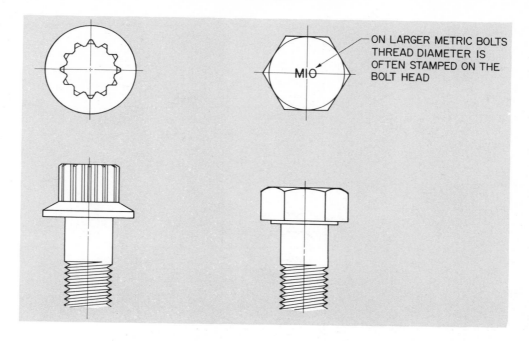

Fig. 19-5. Much study is being done to devise an easy way to identify metric based fasteners from inch based fasteners. The 12-element spline head and imprinted hex head (thread size is stamped on the head) are two methods being considered.

ON LARGER METRIC BOLTS THREAD DIAMETER IS OFTEN STAMPED ON THE BOLT HEAD

M10

DRAWING THREADS

It is time consuming to show threads on the drawing as they would actually appear. For this reason, either the SCHEMATIC or SIMPLIFIED representation of threads generally is used. See the approved methods of representation in Fig. 19-7.

The DETAILED representation (top) looks like the actual screw thread. It is sometimes employed

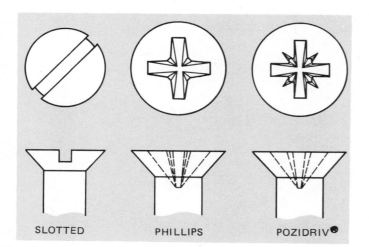

SLOTTED   PHILLIPS   POZIDRIV®

Fig. 19-6. Another method of identification under consideration is the use of the Phillips recessed head on inch based fasteners and the POZIDRIV® recessed head on metric based fasteners.

where confusion might result if the simplified representation were used.

The SCHEMATIC representation (center) is easier to draw. It should not be used for hidden threads or sections of external threads.

The SIMPLIFIED representation of the screw thread is a fast and easy way to draw threads. For this reason, it is widely employed in drafting. It should be avoided where there is a possibility of this representation being confused with other details on the drawing.

Regardless of which thread representation the drafter decides to draw, the thread size must be shown on the drawing. See Fig. 19-8 for the accepted way of presenting thread information.

Threads are understood to be right-hand threads. Left-hand threads are represented by the letters LH.

TYPES OF THREADED FASTENERS

Inch based fasteners of the type required usually can be found in the school shop or a typical hardware store. Metric based fasteners probably will have to be ordered from special suppliers and large hardware stores on special order until they are more widely used. The majority of them are made of steel, brass,

217

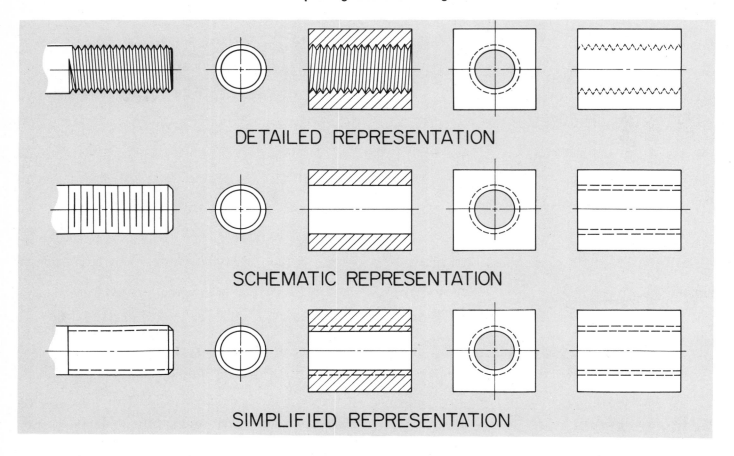

DETAILED REPRESENTATION

SCHEMATIC REPRESENTATION

SIMPLIFIED REPRESENTATION

Fig. 19-7. Approved ways of representing threads on a drawing.

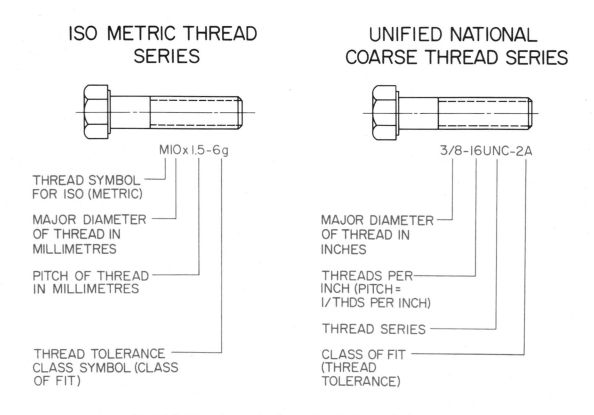

ISO METRIC THREAD
SERIES

MIOx1.5-6g

THREAD SYMBOL
FOR ISO (METRIC)

MAJOR DIAMETER
OF THREAD IN
MILLIMETRES

PITCH OF THREAD
IN MILLIMETRES

THREAD TOLERANCE
CLASS SYMBOL (CLASS
OF FIT)

UNIFIED NATIONAL
COARSE THREAD SERIES

3/8-16UNC-2A

MAJOR DIAMETER
OF THREAD IN
INCHES

THREADS PER
INCH (PITCH=
1/THDS PER INCH)

THREAD SERIES

CLASS OF FIT
(THREAD
TOLERANCE)

Fig. 19-8. How thread size is noted and what each term means.

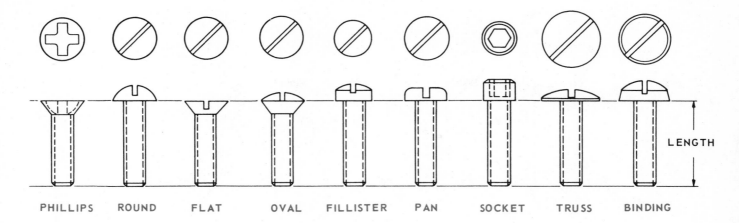

Fig. 19-9. Types of machine screws.

aluminum or nylon. Special applications may require them to be made of other materials.

MACHINE SCREWS

MACHINE SCREWS, Fig. 19-9, are available with single slotted or cross slotted (Phillips and Pozidriv®) heads in round, flat, fillister, pan and oval head styles. Nuts (square and hexagonal) are not furnished with machine screws and must be purchased separately. Machine screws have many applications in general assembly work where small diameter fasteners are needed.

MACHINE BOLTS

MACHINE BOLTS are manufactured with square and hexagonal heads, Fig. 19-10. They are used to assemble products that do not require close tolerance fasteners. Machine bolts are secured by tightening the matching nut.

CAP SCREWS

CAP SCREWS, Fig. 19-11, are used when the assembly requires a stronger, more precise and better appearing fastener. They are primarily employed to bolt two pieces together. The screw passes through a CLEARANCE HOLE in one part and screws into a THREADED HOLE in the other part, Fig. 19-12. They are widely used in assembling machine parts.

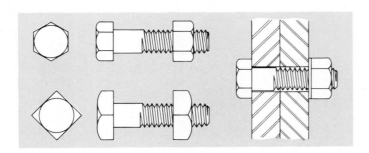

Fig. 19-10. Machine bolts.

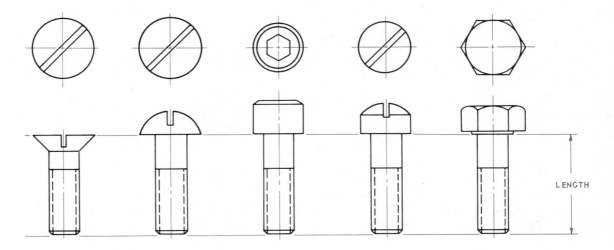

Fig. 19-11. Types of cap screws.

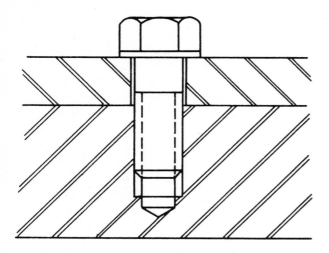

Fig. 19-12. How cap screw works.

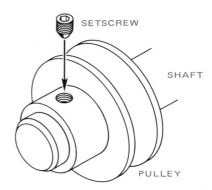

Fig. 19-13. A typical setscrew application.

SETSCREWS

SETSCREWS usually are made of steel and are heat-treated to make them stronger. See Fig. 19-13. Major use is to prevent slippage of pulleys on shafts. Many different head and point styles are available.

STUD BOLTS

STUD BOLTS, Fig. 19-14, are threaded on both ends. One end is threaded into a tapped hole; the

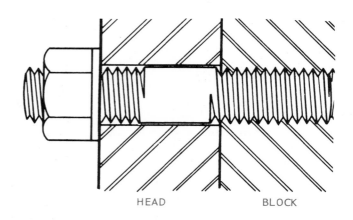

HEAD   BLOCK

Fig. 19-14. Stud bolt.

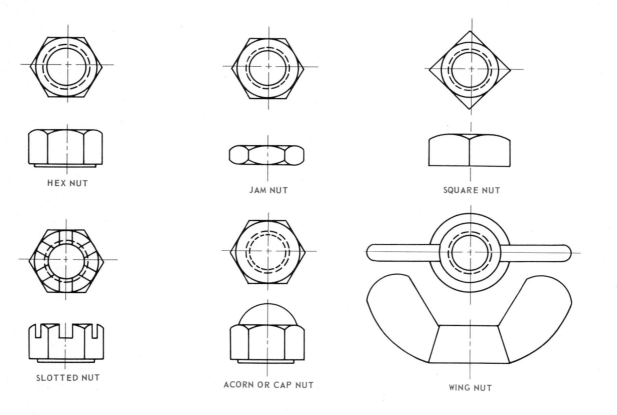

HEX NUT   JAM NUT   SQUARE NUT

SLOTTED NUT   ACORN OR CAP NUT   WING NUT

Fig. 19-15. Common nut styles.

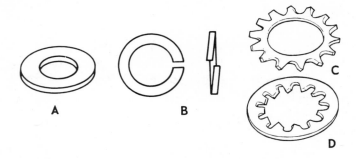

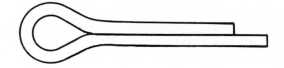

Fig. 19-18. Cotter pin.

Fig. 19-16. Washer types. A—Plain washer. B—Split type lock washer. C—External type lock washer. D—Internal type lock washer.

piece or part to be clamped is fitted over the stud; and the nut is screwed on to clamp the pieces together.

NUTS

NUTS are screwed down on bolts to tighten and hold together the objects through which the bolt passes. A few of the many types available are shown in Fig. 19-15.

WASHERS

WASHERS are used with nuts and bolts to distribute clamping pressure over a larger area, and to prevent the fastener from marring the work surface when it is tightened. Many types are manufactured, Fig. 19-16.

NONTHREADED FASTENERS

Nonthreaded fasteners comprise a large group of holding devices.

RIVETS

Permanent assemblies are made with RIVETS, Fig. 19-17. Two or more pieces of material are held together by these headed pins. Holes are drilled in the

material to be riveted. The shank of the rivet is passed through the hole. After aligning the pieces, the plain end of the rivet is upset or "headed" by hammering to form a second head. The parts are drawn together by the heading process.

Some rivets are expanded by small explosive charges or by other forms of pressure.

COTTER PIN

The COTTER PIN, Fig. 19-18, is fitted into a hole drilled crosswise in a shaft. The ends of the cotter pin are bent down after assembly to prevent parts from slipping or turning off.

KEYS, KEYWAYS AND KEYSEATS

A KEY, Fig. 19-19, is a small piece of metal partially fitted into the shaft and partially in the hub to prevent the rotation of the gear or pulley on the shaft.

The KEYSEAT is machined in the shaft. The KEYWAY is cut in the hub of the mating part.

Different styles of keys have been devised for special applications.

FASTENERS FOR WOOD

The most common fasteners used in wood are nails and screws.

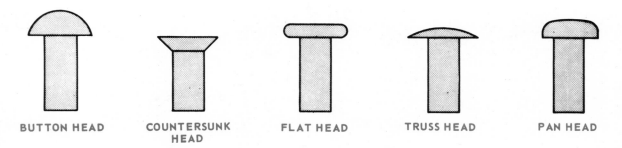

BUTTON HEAD   COUNTERSUNK HEAD   FLAT HEAD   TRUSS HEAD   PAN HEAD

Fig. 19-17. Rivet head styles.

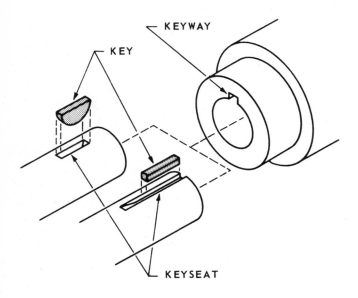

Fig. 19-19. Two types of keys are shown. The half-round key is called a Woodruff key. The other is a rectangular key.

NAILS

Using NAILS is an easy way to fasten wood pieces together. Usually, nails are made of mild steel, but some are made of aluminum. For exterior work, mild steel nails are given a galvanized (zinc) coating.

There will be little, if any, change made when nails are metricated. Nail size is now given as "penny" and is abbreviated with a lower case letter "d." In the future, nail size (length) will be given in millimetres.

WOOD SCREWS

WOOD SCREWS are manufactured from several different kinds of metals. Screw size is indicated by the shank diameter and the length. Head styles and how length is measured is shown in Fig. 19-20.

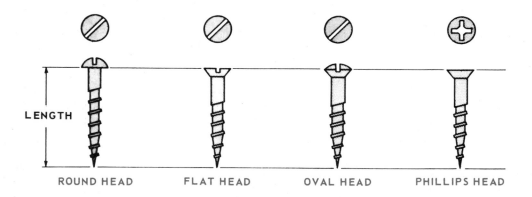

Fig. 19-20. How the various types of wood screws are measured. Screw sizes and head styles will not change when metricated, but their sizes will be expressed in millimetres.

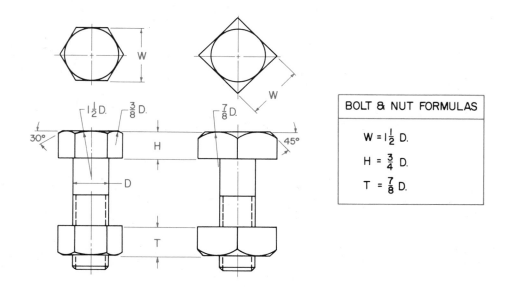

BOLT & NUT FORMULAS
$W = 1\frac{1}{2} D.$
$H = \frac{3}{4} D.$
$T = \frac{7}{8} D.$

Fig. 19-21. Information needed to draw a bolt and nut.

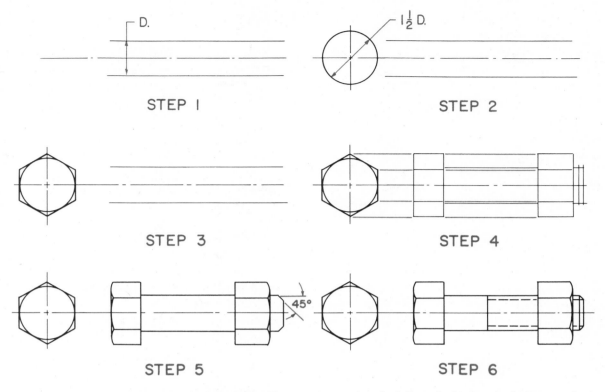

Fig. 19-22. Steps in drawing a bolt and nut. The same procedure is followed whether the bolt or nut is to have a square or hexagonal shaped head.

When wood screws are metricated, screw sizes and head styles will not change. However, their sizes would be expressed in millimetres.

NOTE: Much time can be saved by using (if available) a BOLT HEAD AND NUT TEMPLATE to draw the outline of the bolt head or nut.

HOW TO DRAW BOLTS AND NUTS

The method of drawing nuts and bolts shown in Fig. 19-21 is an approximation, but it is acceptable for most drafting applications. Information needed is:
1. Bolt diameter.
2. Bolt length.
3. Type of head or nut specified.

To draw square and hexagonal bolts and nuts, as in Fig. 19-22, this procedure is recommended:
1. Draw center lines and lines representing the diameter (D).
2. On the center lines, draw a circle (diameter = 1 1/2D).
3. Using a 30-60 degree triangle, circumscribe the hexagon (or square) around the circle.
4. Develop the side view.
5. Draw arcs in bolt head and nut using radii given in Fig. 19-22.
6. Complete by drawing the chamfers on the nut and bolt head. Draw threads (either simplified or schematic) on bolt.

TEST YOUR KNOWLEDGE - UNIT 19

(Write answers on a separate sheet of paper.)

1. Identify 10 different fasteners. Underline the threaded fasteners.
2. What are fasteners?
3. Give five (5) applications that make use of fasteners.
4. List some uses made of screw threads.
   a. _____.
   b. _____.
   c. _____.
   d. _____.
5. What is the difference between a coarse thread and a fine thread on a given size and length bolt?
6. Make sketches of a detailed representation of a screw thread, a schematic representation and a simplified representation.
7. Explain the meaning of the following screw thread size: M10x1.5—6gLH.
8. What fasteners are most commonly used to join wood?

9. Place the letter of the correct definition in the blank to the left of the fastener.

_____Machine screw
_____Cap screw
_____Machine bolt
_____Setscrew
_____Stud bolt
_____Nut
_____Washer

A. Threaded on both ends.
B. Fitted on a bolt.
C. Distributes the clamping pressure of the nut or bolt.
D. Prevents slippage of a pulley on a shaft.
E. Used for general assembly work where small diameter fasteners are needed.
F. Used to assemble products that do not require close tolerance fasteners.
G. Used when an assembly requires a stronger, more precise and better appearing fastener.

OUTSIDE ACTIVITIES

1. Draw M16 x 2 hexagonal and square head bolts and nuts on the same sheet. The bolts are 100 mm long. Allow 75 mm between the drawings. Use a simplified thread representation.
2. Draw M24 x 3 hexagonal and square head bolts and nuts on the same sheet. The bolts are 75 mm long. Allow 75 mm between the drawings. Use a schematic thread representation.
3. Secure samples of metric based machine screws, cap screws, machine bolts, setscrews, stud bolts, nuts and washers. Mount them on a display board so they can be used in the drafting room. Hint: Metric fasteners are used on most foreign made automobiles and motorcycles.
4. Get an assembly showing a key, keyseat and keyway in use.
5. Secure an example of a metric setscrew application.

Unit 20
ELECTRICAL AND
ELECTRONICS DRAFTING

Our world, as we know it today, could not exist without electricity and the electronic devices it powers. You will realize how true this is by just listing the electrical devices you depend upon in your home — TV set, transistor radio, washer, refrigerator, stereo, range and the lighting to name but a few.

Other electronic devices which may not be as familiar are the computers that aid the astronauts in their flights to the moon and the devices that control the machines that make, inspect, assemble and test so many of the products we use. See Fig. 20-1. Many hobbies are electronically oriented, Fig. 20-2.

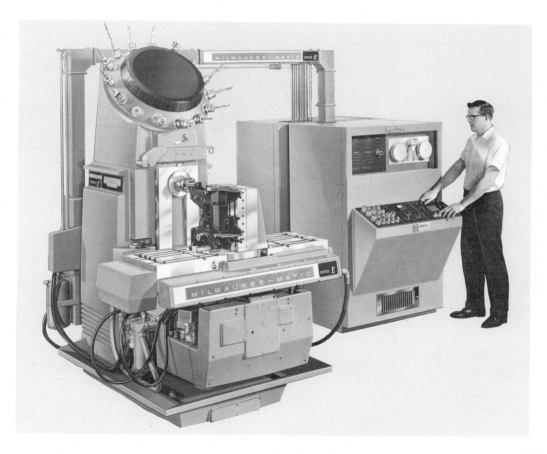

Fig. 20-1. Modern machine tools, like this machining center, are controlled electronically. Note the various cutting tools.   (Kearney and Trecker)

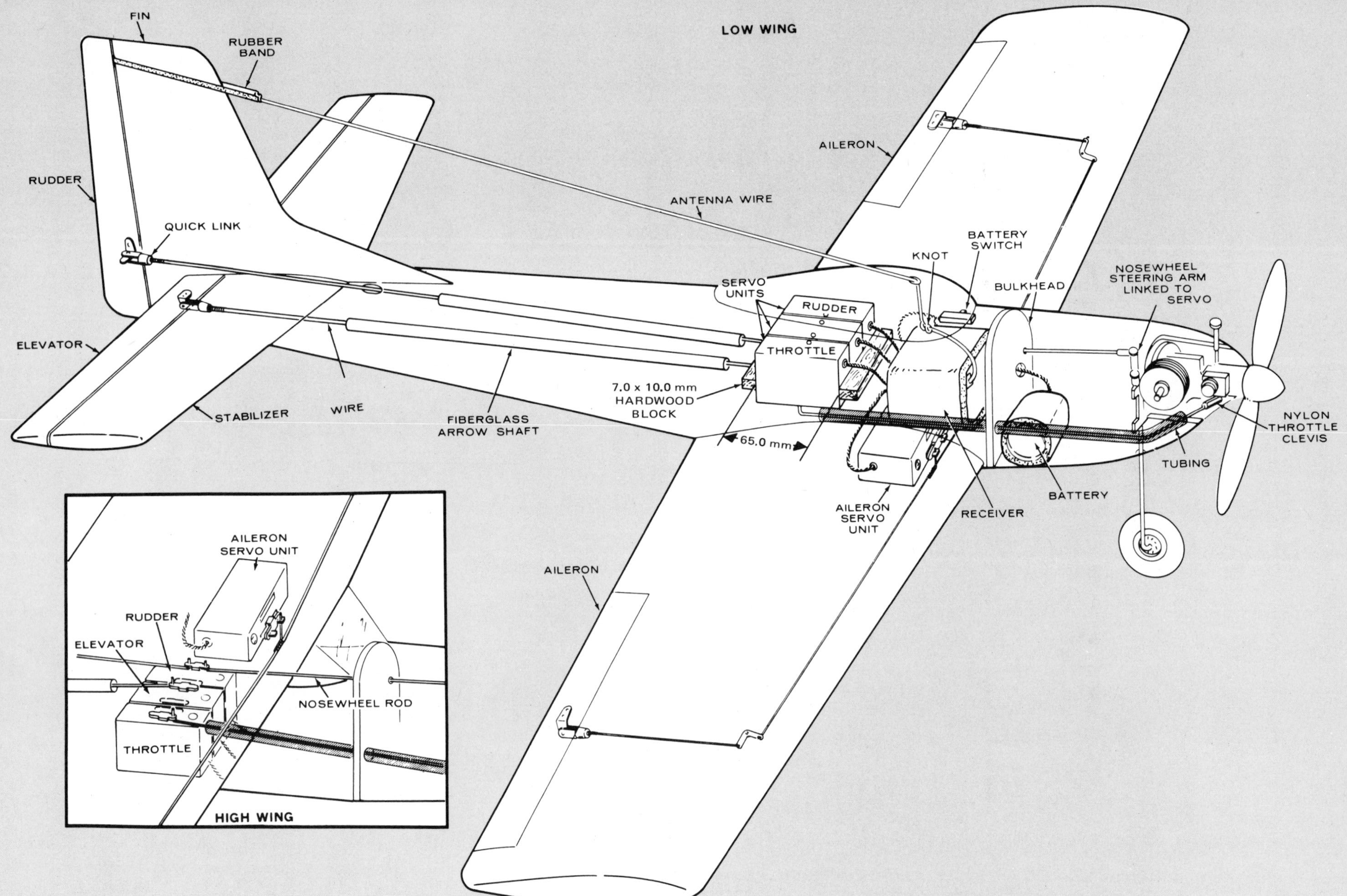

NYLON THROTTLE CLEVIS

NOSEWHEEL STEERING ARM LINKED TO SERVO

TUBING

BATTERY

BULKHEAD

BATTERY SWITCH

KNOT

RECEIVER

AILERON SERVO UNIT

AILERON

LOW WING

ANTENNA WIRE

SERVO UNITS

RUDDER

THROTTLE

65.0 mm

7.0 x 10.0 mm HARDWOOD BLOCK

AILERON

FIBERGLASS ARROW SHAFT

WIRE

RUBBER BAND

QUICK LINK

STABILIZER

FIN

RUDDER

ELEVATOR

AILERON SERVO UNIT

RUDDER

ELEVATOR

NOSEWHEEL ROD

THROTTLE

HIGH WING

Fig. 20-2. Radio control in a model airplane. Many hobbies are electronically oriented. (Heath Company)

226

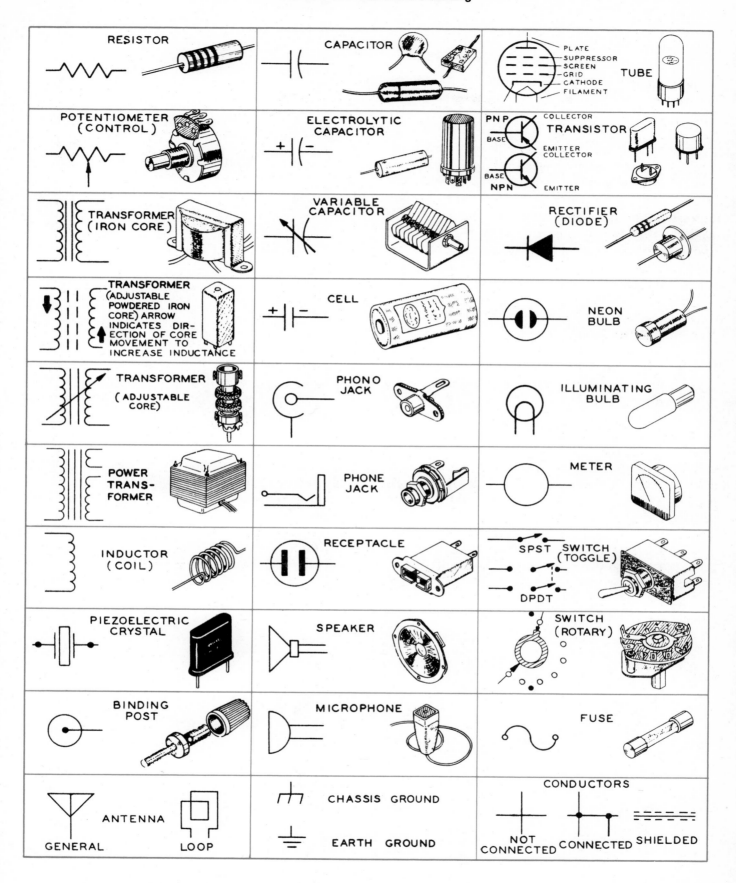

Fig. 20-3. Typical electronic symbols and related illustrations.

ELECTRICAL AND ELECTRONIC DRAFTING

Electrical and electronics drafting is done in much the same manner as conventional drafting. The same type equipment is needed.

However, instead of using regular multiview drawings, a large portion of electrical and electronics drafting is diagrammatic in character. Considerable use is made of symbols, Fig. 20-3. The symbols represent the various components and wires that make up the electronic circuit. They are easier and quicker to draw than the actual part. Symbols are combined on a diagram that shows the function and relation of each component in the circuit.

TYPES OF DIAGRAMS

The drafter working in electrical and electronics drafting must be familiar with the following different types of diagrams:

A SCHEMATIC DIAGRAM, Fig. 20-4, is a drawing using symbols and single lines to show the electrical connections and functions of a specific electronic circuit. The various components that make up the circuit are drawn without regard to their actual physical size, shape or location.

A CONNECTION OR WIRING DIAGRAM, Fig. 20-5, is commonly used to show the distribution of electricity on architectural drawings. This type of

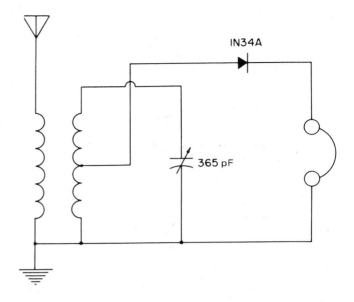

Fig. 20-4. A schematic diagram of a crystal radio.

diagram also may be drawn to show the general physical arrangement of the transistors, diodes, resistors, switches, etc., that make up an electronic circuit. See Fig. 20-6.

A BLOCK DIAGRAM, Fig. 20-7, is a simplified way to show the operation of an electronic device. This diagram utilizes "blocks" (squares and/or rectangles) that are joined by a single line. It reads from left to right.

A PICTORIAL DIAGRAM, Fig. 20-8, shows the components in pictorial form and in their proper location. The pictorial diagram is used extensively by

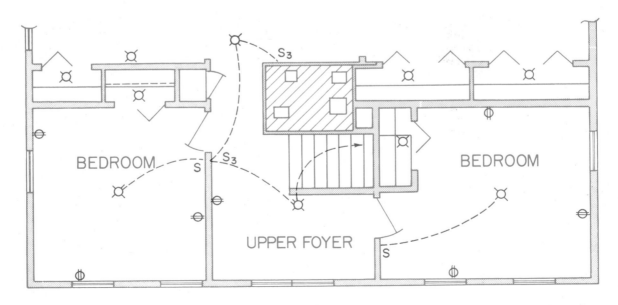

Fig. 20-5. A wiring diagram showing the location of switches, plugs and lighting in a modern home.

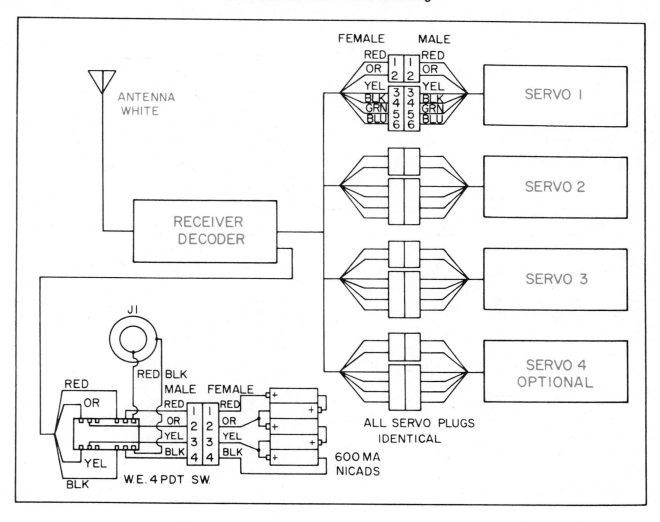

Fig. 20-6. A wiring diagram of a radio control installation similar to the unit shown in Fig. 20-2.

electronic kit manufacturers because it is so easy to understand.

DRAWING SYMBOLS

Symbols in electrical and electronic circuit diagrams need not be drawn to any particular scale.

However, they should be shaped correctly, drawn large enough to be seen clearly and kept in proportion. See Fig. 20-9.

Symbols and lines are drawn the same weight as a visible object line. A darker line may be used when a portion of the diagram must be emphasized.

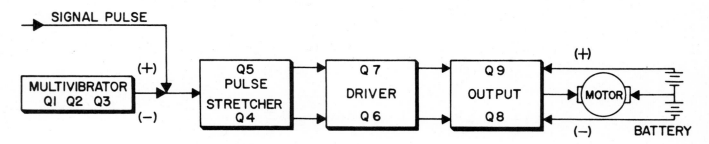

Fig. 20-7. A block diagram of servo that may be used to activate the controls of a radio controlled model airplane, boat or car.

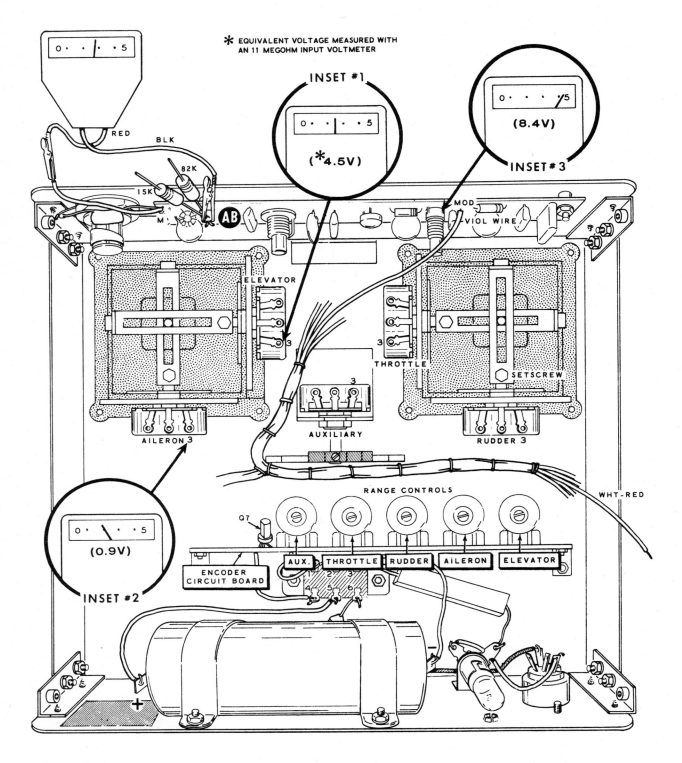

* EQUIVALENT VOLTAGE MEASURED WITH
AN 11 MEGOHM INPUT VOLTMETER

INSET #1

(*4.5V)

(8.4V)

INSET #3

RED

BLK

82K

15K

M

AB

MOD

VIOL WIRE

ELEVATOR

THROTTLE

SETSCREW

AILERON 3

AUXILIARY

RUDDER 3

(0.9V)

INSET #2

RANGE CONTROLS

WHT-RED

Q7

ENCODER
CIRCUIT BOARD

AUX. THROTTLE RUDDER AILERON ELEVATOR

+

Fig. 20-8. A pictorial diagram in which electrical components are drawn in pictorial form.   (Heath Company)

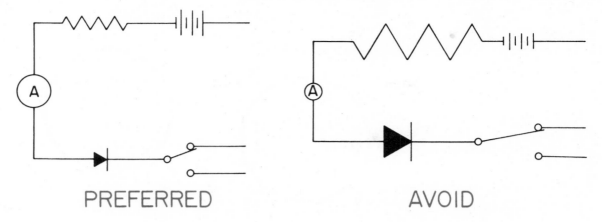

PREFERRED   AVOID

Fig. 20-9. Draw symbols in proportion.

TEST YOUR KNOWLEDGE - UNIT 20

(Write answers on a separate sheet of paper.)

1. How does electrical and electronics drafting differ from conventional drafting?
2. Why are symbols used to represent electrical and electronic components on drawings?
3. A SCHEMATIC DIAGRAM is a drawing _____.
4. How does a BLOCK DIAGRAM differ from a WIRING DIAGRAM?
5. What is a PICTORIAL DIAGRAM?
6. Symbols and lines used in electrical and electronics drafting are drawn the same weight as a _____ _____ line.
7. The electrical diagrams used on house plans are called _____ diagrams.

OUTSIDE ACTIVITIES

1. Make a schematic diagram of a desk lamp.
2. Draw the following electronic symbols (space them uniformly on the sheet):

a. battery (single cell)	k. capacitor (variable)
b. battery (three cell)	l. line plug
c. antenna	m. loud speaker
d. crystal	n. microphone
e. crystal diode	o. variable resistor
f. fuse	p. switch
g. ground	q. transistor
h. headphones	r. resistor
i. jack	s. wires connected
j. lamp	t. wires not connected

3. Prepare a schematic diagram of a two cell flashlight.
4. Make a wiring diagram of your bedroom.
5. Secure a small battery powered toy and examine how it operates. Make a suitable diagram showing your findings.
6. Prepare a pictorial diagram of a crystal radio set.
7. Make a schematic diagram of a single tube radio receiver.
8. Make a suitable diagram showing four batteries, switch and lamp wired in series.
9. Make a suitable diagram showing four batteries, switch and lamp wired in parallel.

Unit 21
ARCHITECTURAL DRAFTING

The building industry probably will be one of the last business enterprises to go metric. However, provisions are being made for the changeover. Lumber producers have decided upon a "soft" conversion of existing lumber sizes. This will not require a basic redimensioning of lumber sizes as would the SI or "hard" conversion. A "soft" conversion simply means that inches have been changes to millimetres, pounds to kilograms, etc. A "hard" conversion with metric engineering standards is one in which all of the designing is done in preferred metric sizes.

At present, some plumbing fixtures made to SI standards are being imported. Problems have been encountered trying to join these fixtures to inch based pipes and fittings.

In the United States, most buildings are designed to a 4-inch base module. See Fig. 21-1. All lumber, blocks, bricks, panel stock (plywood, hardboard, etc.) and components such as windows and doors are specified in multiples of the 4 in. module. Windows, for example, usually are 2'-8" or 3'-0" wide.

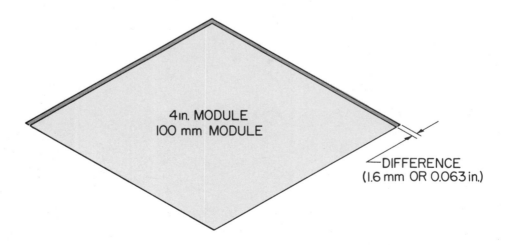

4in. MODULE
100 mm MODULE

DIFFERENCE
(1.6 mm OR 0.063 in.)

Fig. 21-1. Most buildings and building materials now used are designed to multiples of the basic 4 in. module. A 100 mm x 100 mm module is the metric based unit being proposed.

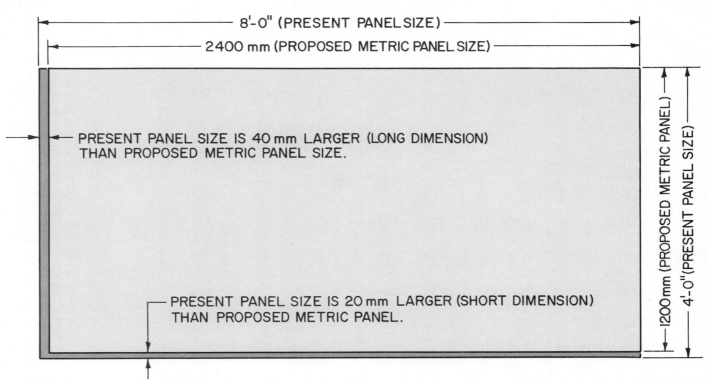

Fig. 21-2. A comparison of the conventional size panel and the proposed metric size panel. The proposed metric size panel would be too small to use for a replacement for the conventional 4 ft. x 8 ft. panel.

A 100 mm by 100 mm module, Fig. 21-2, is being recommended as the basic metric module. All metric based building materials and components will be based on multiples of this module.

A 100 mm by 100 mm module is almost, BUT NOT QUITE, the same size as the customary 4 in. by 4 in. module. The difference is small, but is enough to cause major problems if ALL materials used in constructing a building are not designed to the same basic module.

ARCHITECTURAL DRAFTING

There is little difference between the metric and customary measuring systems in architectural draft-

METRIC SCALE	USE	CUSTOMARY EQUIVALENT
1:10	Construction Details	1'' = 1'—0'' (1:12)
1:20	Construction Details	3/4'' = 1'—0'' (1:16)
1:25	Construction Details	1/2'' = 1'—0'' (1:24)
1:50	Plans, Elevations	1/4'' = 1'—0'' (1:48)
1:100	Plot Plan	1/8'' = 1'—0'' (1:96)
1:200	Plot Plan	1/16'' = 1'—0'' (1:192)
1:500	Site Plan	1/32'' = 1'—0'' (1:384)

Fig. 21-3. A few of the metric scales recommended for use in metric based architectural drafting.

ing. Dimensions will be given in millimetres and metres instead of inches and feet.

The ISO "A" series of drawing sheet sizes will be used. Scales for use in metric based architectural drafting are shown in Fig. 21-3. All structures will be designed and constructed on the 100 mm module.

BUILDING A HOME

Plans which give the craftworker the information needed to construct a building in which people may live, work or play are called ARCHITECTURAL DRAWINGS. See Fig. 21-4.

Buying a home is probably the largest investment you will make in your lifetime. An understanding of the basic principles of architectural drafting will be a great help. This is especially true if you plan to design or construct a new home, modernize an existing home or judge the value of a home offered for sale.

The ability to read and interpret architectural drawings is essential to those who work in the construction industry. These workers include carpenters, masons, plumbers, electricians, roofers, etc. It is also useful to workers in lumber yards, hardware and building supply stores.

Fig. 21-4. Typical small home. Plans for this home will be given in this Unit.
(National Plan Service, Inc.)

To make sure your own home is built to your specific requirements, it is important that you have a good plan and a well defined contract with your builder.

Plans provide the vast amount of information needed to construct a modern home. They should incorporate applicable aspects of the local and state building codes.

BUILDING CODES

Building codes are laws which provide for the health, safety and general welfare of the people in the community.

Building codes are based on standards developed by government and private agencies.

BUILDING PERMIT

In most communities, the building contractor or owner must file a formal application for a building permit. Plans and specifications are submitted for the proposed structure. They are reviewed by building officials to determine whether they meet local building code requirements.

Work is inspected by local building officials as construction progresses. An on-site inspection card is signed as each phase of construction is approved.

PLANS

Because it is not possible to include all construction details on a single sheet, a set of typical house plans will include a plot plan, foundation and/or basement plan, floor plans, elevations (front, rear and side views of the home), wall sections, and built-in cabinet and fireplace details.

SCALE

The plans generally are drawn to a scale of 1:50. This means that 1 mm on the drawing equals 50 mm in the building being constructed.

A larger scale, 1:10, is often used when greater detail is needed on a structural part. Framing plans are often drawn 1:100.

PLOT PLAN

The PLOT PLAN, Fig. 21-5, shows the location of the building on the building site. Also shown are

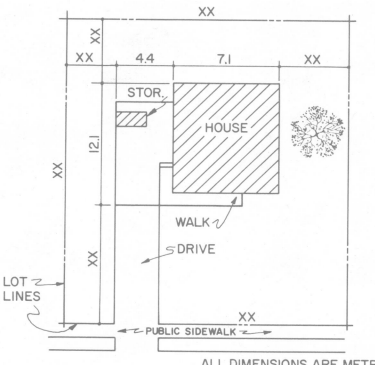

Fig. 21-5. Plot plan for the house shown in Fig. 21-4.

walks, driveways and patios. Overall building and lot dimensions are included. Contour lines (which indicate the slope of the land) are sometimes shown.

ELEVATIONS

ELEVATIONS are the front, rear and side views of the house, Fig. 21-6. They are made of lines which are visible when the building is viewed from various positions. Elevations include floor levels, grade lines, window and door heights, roof slope and the types of materials to be used on the walls and roof.

Foundation and footing lines below grade are indicated by hidden lines. Only visible edges are shown above grade level.

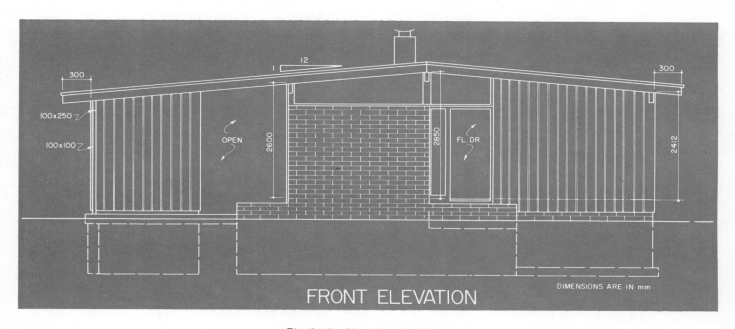

Fig. 21-6. Elevation drawing.

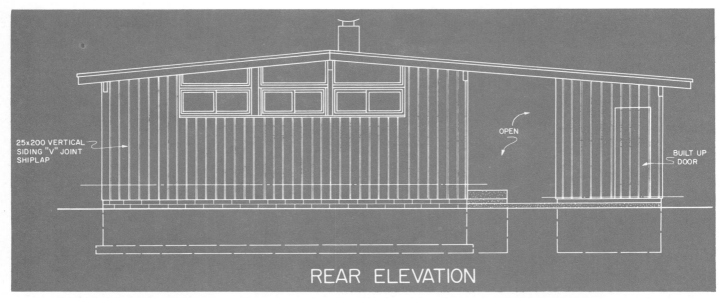

REAR ELEVATION

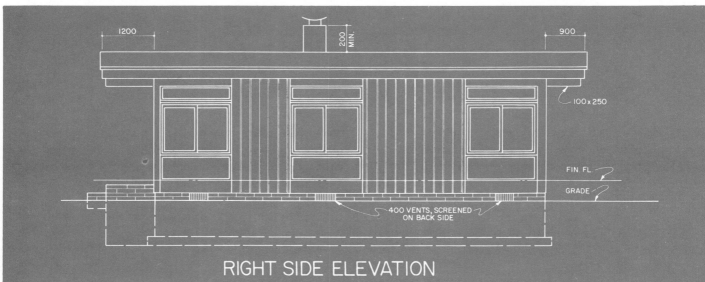

RIGHT SIDE ELEVATION

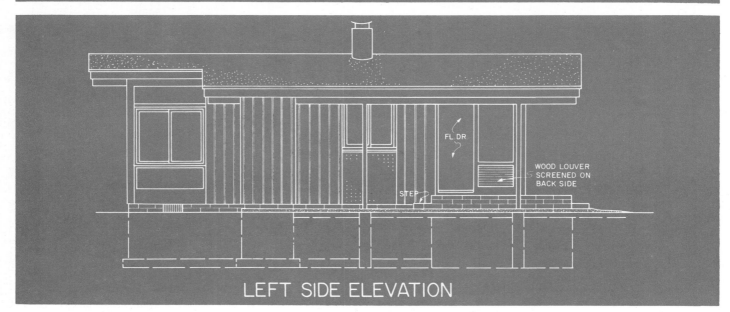

LEFT SIDE ELEVATION

Fig. 21-6. (Continued).

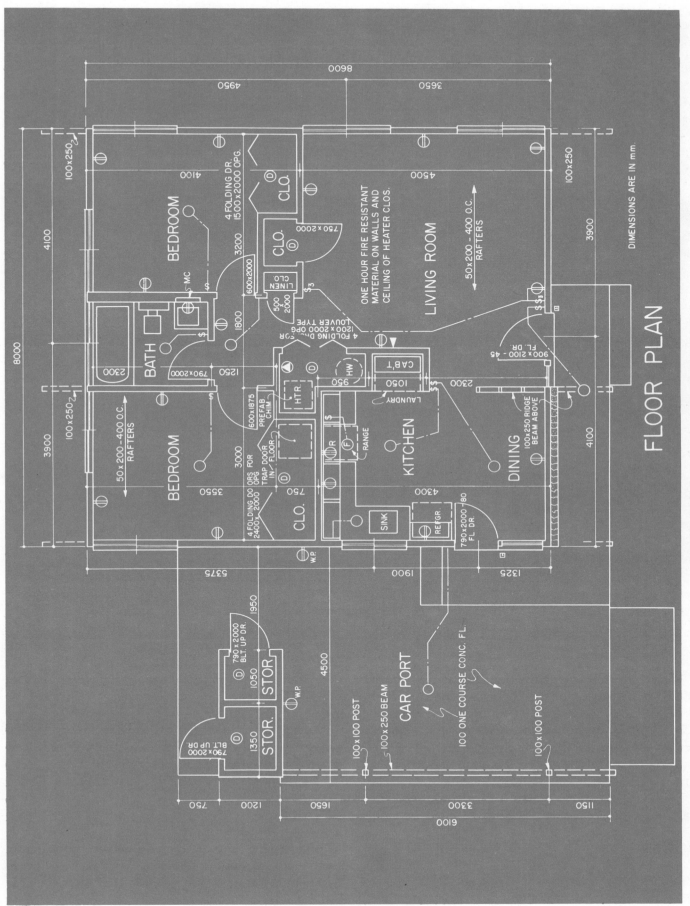

FLOOR PLAN

Fig. 21-7. Floor plan. (National Plan Service, Inc.)

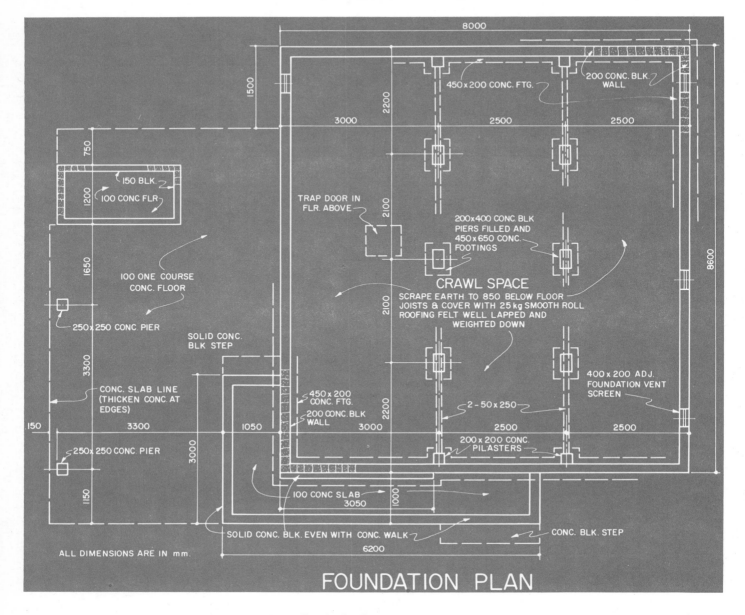

Fig. 21-8. Foundation plan.

FLOOR PLANS

The size and shape of the building, as well as the interior arrangement of the rooms are shown on the FLOOR PLAN. See Fig. 21-7. Additional information such as the location and sizes of the interior partitions, doors, windows, stairs and utility installations (plumbing, electrical, etc.) is also included.

Foundation and basement plans frequently are combined on a single sheet. See Fig. 21-8.

SECTIONS

SECTIONAL VIEWS, Fig. 21-9, are used to give construction details of the structure from basement or foundation footing to the ridge of the roof. "Break lines" are often incorporated into the section. This reduces the size of the drawing and saves time in drawing the plan.

Along with the sizes of framing materials, the types and kinds of sheathing, insulation, interior and exterior wall surfaces, etc., are indicated.

DETAILS

Information to assist craftworkers in constructing such things as built-in cabinets, fireplaces and other pertinent information are given on DETAIL SHEETS. See Fig. 21-10.

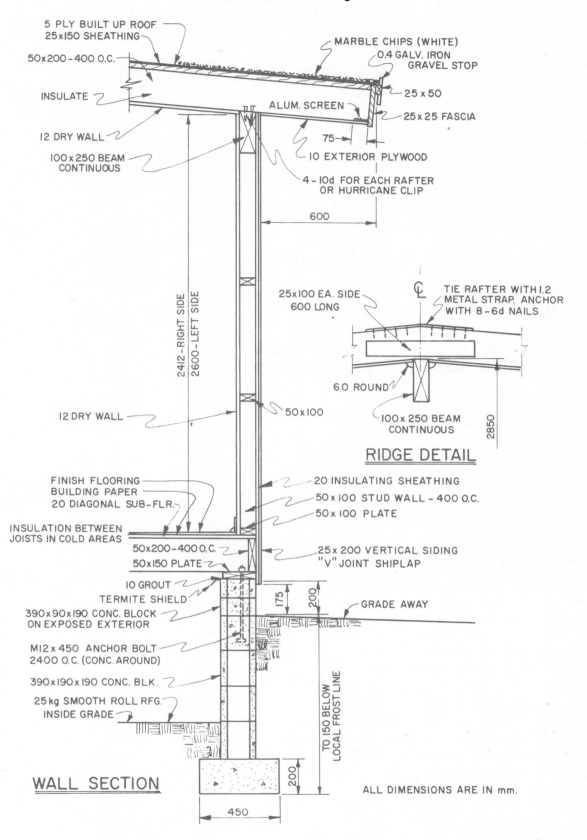

5 PLY BUILT UP ROOF
25x150 SHEATHING

50x200 - 400 O.C.

MARBLE CHIPS (WHITE)

0.4 GALV. IRON
GRAVEL STOP

INSULATE

25 x 50

12 DRY WALL

ALUM. SCREEN

25 x 25 FASCIA

100 x 250 BEAM
CONTINUOUS

75

10 EXTERIOR PLYWOOD

4 - 10d FOR EACH RAFTER
OR HURRICANE CLIP

600

2412 - RIGHT SIDE
2600 - LEFT SIDE

25x100 EA. SIDE
600 LONG

TIE RAFTER WITH 1.2
METAL STRAP. ANCHOR
WITH 8 - 6d NAILS

CL

12 DRY WALL

50 x 100

6.0 ROUND

100 x 250 BEAM
CONTINUOUS

2850

RIDGE DETAIL

FINISH FLOORING
BUILDING PAPER
20 DIAGONAL SUB-FLR.

20 INSULATING SHEATHING
50 x 100 STUD WALL - 400 O.C.
50 x 100 PLATE

INSULATION BETWEEN
JOISTS IN COLD AREAS

50x200 - 400 O.C.

50x150 PLATE

25 x 200 VERTICAL SIDING
"V" JOINT SHIPLAP

10 GROUT
TERMITE SHIELD

175

200

GRADE AWAY

390x90x190 CONC. BLOCK
ON EXPOSED EXTERIOR

M12 x 450 ANCHOR BOLT
2400 O.C. (CONC. AROUND)

390x190x190 CONC. BLK.

25 kg SMOOTH ROLL RFG.
INSIDE GRADE

TO 150 BELOW
LOCAL FROST LINE

WALL SECTION

200

450

ALL DIMENSIONS ARE IN mm.

Fig. 21-9. Wall section.

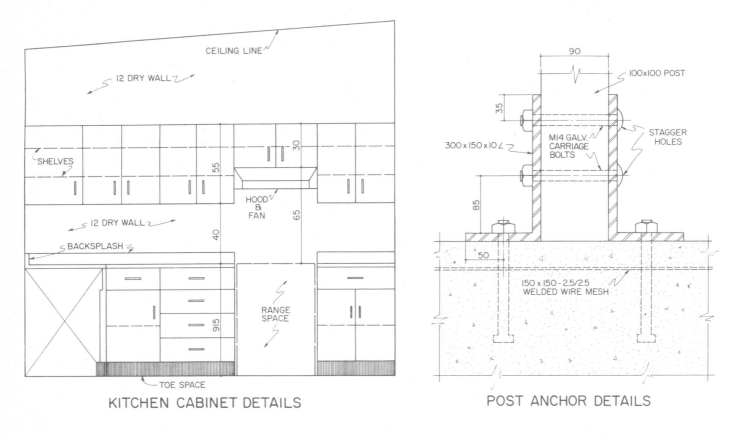

KITCHEN CABINET DETAILS

POST ANCHOR DETAILS

Fig. 21-10. Details found on a typical architectural drawing.

ARCHITECTURAL DRAFTING TECHNIQUES

In general, most conventional drafting techniques apply to architectural drafting.

SYMBOLS, "SPEC" SHEETS

Wide use of SYMBOLS, Fig. 21-11, will be noted in architectural drafting. Symbols are employed on the plans because it is not practical to show items such as doors, windows, plumbing fixtures, etc., as they would actually appear in the structure.

Sheets of specifications usually accompany house plans. These "spec sheets" describe in writing things that cannot be easily shown on the drawings. Things such as quality of materials, how installation of specific items are to be made, etc.

SCALE

With few exceptions, architectural plans are drawn to scale. When drawn to scale, the views are shown in accurate proportion to the full size structure. A 1:50 scale would mean that 1 mm on the drawing equals 50 mm in the structure being built.

There is no metric architect's scale. A metric rule of the proper scale ratio is all that is required. See Fig. 21-12.

TITLE BLOCK

A TITLE BLOCK, Fig. 21-13, for architectural drawings should include the following:
1. Type of structure (house, garage, etc.).
2. Where the structure is to be located.
3. Architect's name and/or drafting personnel making drawing.
4. Date.
5. Drawing scale or scales being used.
6. Sheet number and number of sheets making up the full set of drawings (Sheet 1 of 10, etc.).

The title block may be made any convenient size. However, 50 mm by 100 mm is a suitable size for most home plans. Usually, it is located in the lower right corner of the drawing sheet.

LETTERING

Conventional single stroke Gothic lettering may be used on architectural drawings. However, most archi-

PLAN SYMBOLS

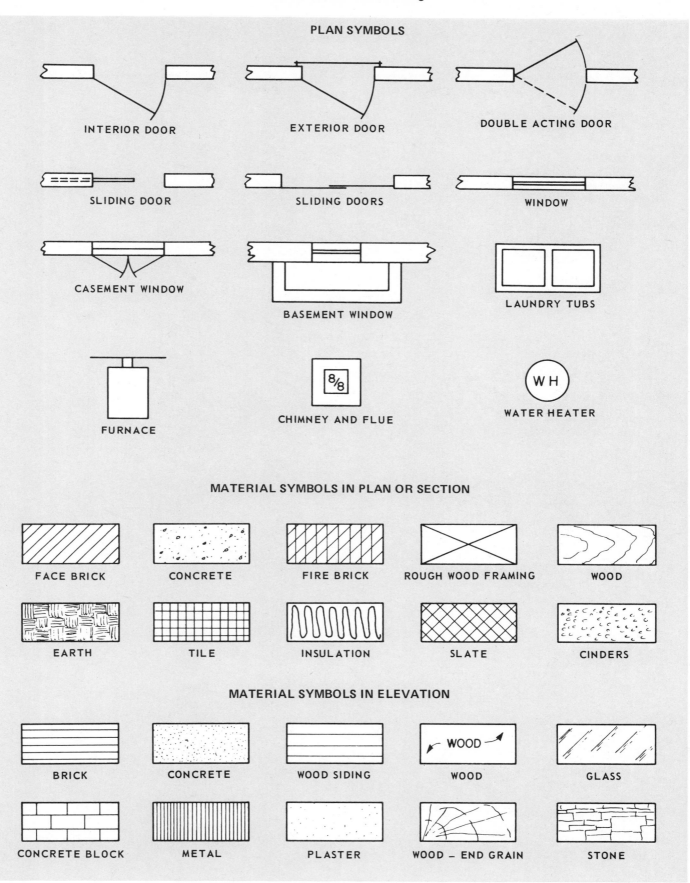

MATERIAL SYMBOLS IN PLAN OR SECTION

MATERIAL SYMBOLS IN ELEVATION

Fig. 21-11. Symbols used to indicate materials and fixtures on drawings.

KITCHEN PLAN SYMBOLS

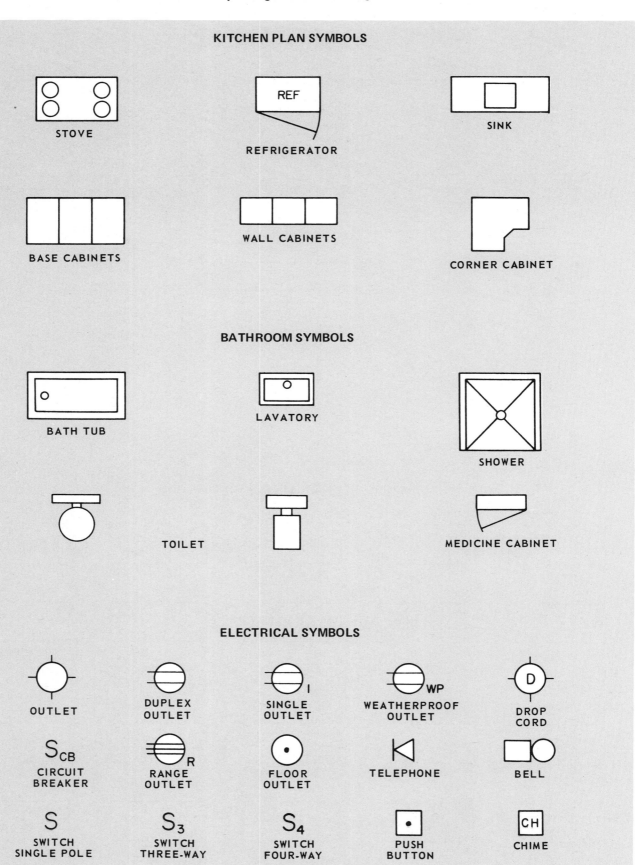

STOVE

REF

REFRIGERATOR

SINK

BASE CABINETS

WALL CABINETS

CORNER CABINET

BATHROOM SYMBOLS

BATH TUB

LAVATORY

SHOWER

TOILET

MEDICINE CABINET

ELECTRICAL SYMBOLS

OUTLET

DUPLEX OUTLET

SINGLE OUTLET

WEATHERPROOF OUTLET

DROP CORD

S_{CB}
CIRCUIT BREAKER

RANGE OUTLET

FLOOR OUTLET

TELEPHONE

BELL

S
SWITCH SINGLE POLE

S_3
SWITCH THREE-WAY

S_4
SWITCH FOUR-WAY

PUSH BUTTON

CHIME

Fig. 21-11. (Continued).

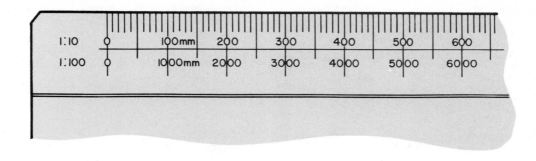

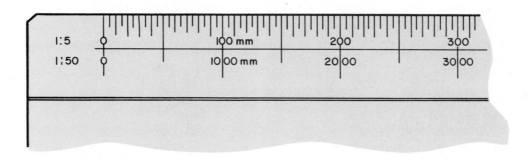

Fig. 21-12. Two metric scales used in architectural drafting.

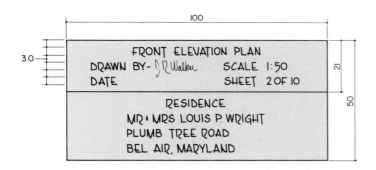

Fig. 21-13. Typical title block.

tectural lettering follows the Roman alphabet of letters and numbers. See Fig. 21-14. Many architects and architectural drafting personnel develop their own distinctive style of lettering. Whether you decide to use the single stroke Gothic lettering or develop a style of your own, remember that your lettering must be legible.

DIMENSIONING

There is only a slight difference between dimensioning techniques used on architectural drawings and on other forms of drafting. The dimensions are read from the bottom and right sides of the drawing sheet. Dimensions may be placed on all sides of the drawing and through the drawing itself. Dimension figures, usually 3 or 4 mm high, are written above the dimension line. The dimension line may be capped with arrowheads or small darkened circles, Fig. 21-15. Observe the following dimensioning rules.

ABCDEFGHIJKLMNOPQRST
UVWXYZ & 1234567890
abcdefghijklmnopqrstuvwxyz-

Fig. 21-14. One style of architectural lettering.

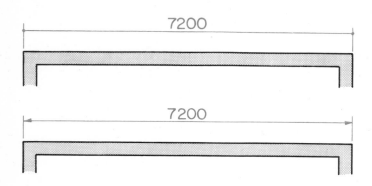

Fig. 21-15. Ends of dimension lines may be capped by arrowheads, or small darkened circles may be used.

DIMENSIONING RULES

1. Make sure that the dimensioning is complete. A builder should never have to assume or measure a distance on the drawing for a dimension.
2. ALWAYS letter dimensions full-size regardless of the drawing scale.
3. Dimensions and notes should be at least 6 mm away from the view.
4. Align the dimensions across the view wherever possible.
5. Place overall dimensions outside the view.
6. Overall dimensions are determined by adding detail dimensions. Do not scale the drawing for these dimensions.
7. Dimension partitions center to center or from center to outside wall.
8. Room size may be indicated by stating the width and length of the room.
9. Windows, doors, beams, etc., are dimensioned to their centers.
10. Window and door sizes and types are given in a SCHEDULE, Fig. 21-16.

ARCHITECTURAL ABBREVIATIONS

Time and space can be saved when lettering architectural drawings by using abbreviations. Some that are commonly used follow:

Asphalt Tile - AT	Elevation - EL
Beam - BM	Exterior - EXT
Bedroom - BR	Floor - FLR
Brick - BRK	Footing - FTG
Ceiling - CLG	Grade Line - GL
Center Line - CL or ₵	Lavatory - LAV
Concrete - CONC	Living Room - LR
Drawing - DWG	Plaster - PL
Door - DR	Room - RM

OBTAINING INFORMATION

Students studying architectural drawing need to do a great deal of research into catalogs and reference books. This is the only way that the many standard sizes essential to completing an accurate set of plans can be found.

Additional information may be found in mail order catalogs, lumber and building supply company literature, from the Federal Housing Administration (FHA) and copies of your local building ordinances. Specialized books in architecture, carpentry, plumbing and other building areas are excellent sources of information.

PLANNING A HOME

When planning a home, you must determine how much money will be available for building purposes. It is suggested that a home should not cost more than four or five times your yearly TAKE HOME salary.

WINDOW SCHEDULE				
MARK	REQ'D	OVERALL SASH SIZE WIDTH	HEIGHT	DESCRIPTION
W-1	4	1600	x 1200	ALUMINUM SLIDING WINDOW WITH INSULATING SASH, FIBERGLAS SCREENS
W-2	3	1600	x 600	SAME AS ABOVE
W-3	1	1200	x 900	"
W-4	1	800	x 1700	6.0 PLATE GLASS

Fig. 21-16. Window schedule of the type found on architectural drawings.

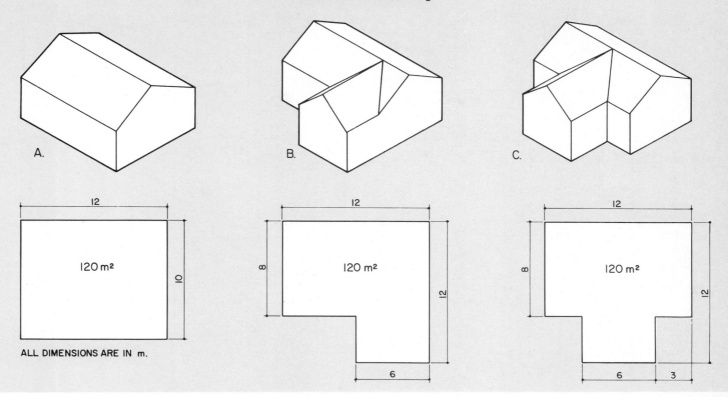

ALL DIMENSIONS ARE IN m.

Fig. 21-17. Examples of floor areas of 120 m². Plans shown in "B" and "C" are slightly more costly to construct because the extra corners require more material and labor to complete.

One way to determine home size is by the square metre (m²) method. To set up a typical planning problem, assume that $42,000.00 is available to build the home.

Building costs vary from one area of the country to another area. A local contractor can give you the approximate cost PER SQUARE FOOT. You will have to translate this to cost PER SQUARE METRE. Assume the approximate cost of frame construction is $350.00 per square metre (m²).

Divide the cost per m² ($350.00) into the money available to build the house ($42,000.00). You will find that a frame house of 120 m² of floor area can be built. Outlines of homes having 120 m² of floor area are shown in Fig. 21-17.

After determining the approximate size and shape of the proposed house, you should try to develop a suitable floor plan. Room use and space available must be carefully considered and balanced.

Be sure to include closet and storage facilities. Using scale cutouts of the furniture that you plan to use in the various rooms will help you in your planning. The cutouts MUST be made to the same scale as the floor plan. Prepare a scale drawing of the room showing all openings AND THE AREAS THE DOORS TAKE UP WHEN OPENED.

By moving the cutouts, it is easy to find whether the room is large enough for the intended use.

Cutout planning for a typical room is shown in Fig. 21-18. Sizes of some furniture and appliances are given in Fig. 21-19. Dimensions of items not shown may be found in manufacturer's catalogs, home magazines and mail order catalogs.

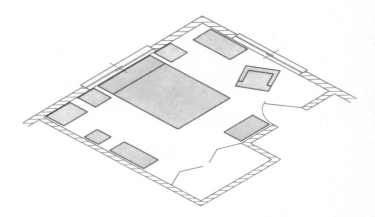

Fig. 21-18. How cutouts may be used to plan a bedroom.

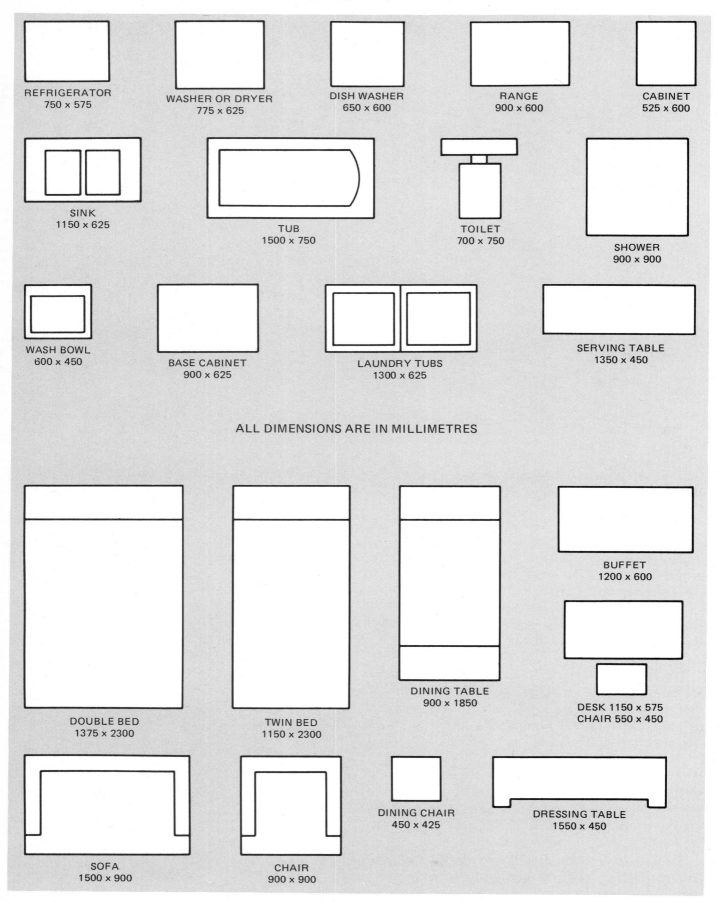

REFRIGERATOR
750 x 575

WASHER OR DRYER
775 x 625

DISH WASHER
650 x 600

RANGE
900 x 600

CABINET
525 x 600

SINK
1150 x 625

TUB
1500 x 750

TOILET
700 x 750

SHOWER
900 x 900

WASH BOWL
600 x 450

BASE CABINET
900 x 625

LAUNDRY TUBS
1300 x 625

SERVING TABLE
1350 x 450

ALL DIMENSIONS ARE IN MILLIMETRES

BUFFET
1200 x 600

DESK 1150 x 575
CHAIR 550 x 450

DINING TABLE
900 x 1850

DOUBLE BED
1375 x 2300

TWIN BED
1150 x 2300

DINING CHAIR
450 x 425

DRESSING TABLE
1550 x 450

SOFA
1500 x 900

CHAIR
900 x 900

Fig. 21-19. Furniture dimensions that will aid in planning a home.

TEST YOUR KNOWLEDGE - UNIT 21

(Write answers on a separate sheet of paper.)

1. Architectural drawings are _____.
2. The ability to read and interpret architectural drawings is essential to workers such as:
   a. _____.
   b. _____.
   c. _____.
   d. _____.
   e. _____.
3. What are building codes?
4. Architectural drawings are generally drawn to a scale of _____.
5. The location of the buildings on the building sites is shown on the _____. This drawing also shows the walks, driveways and patios.
6. Drawings that show the various views (front, rear and side views) of the structure are called _____.
7. The floor plan shows:
   a. Size and shape of the rooms.
   b. Location and sizes of the windows and doors.
   c. Location of the utilities (plumbing, heating and electrical fixtures).
   d. All of the above.
   e. None of the above.
8. Construction details of the structure from basement or foundation to the ridge of the roof are shown on _____.
9. Make a sketch showing two recommended methods of dimensioning architectural drawings.

OUTSIDE ACTIVITIES

1. Make a scale drawing (1:50) of the school drafting room. Use cutouts to determine an efficient layout for the room.
2. Prepare a drawing of your bedroom, showing the location of the furniture in the room. Use a scale of 1:25.
3. Draw the floor plan of a two-car garage with a workshop at one end.
4. Design and draw a full set of plans for a two room summer cottage or hunting cabin.
5. Draw the plans necessary to convert your basement into a workshop or recreation room.
6. Design and draw the plans necessary to construct a storage shed.
7. Design a small two bedroom house that can be constructed at minimum cost. Construct a scale model of it.
8. Plan a structure to house a sports car sales agency. Construct a scale model of your design.

Unit 22
AUTOMATED DRAFTING

COMPUTER GRAPHICS is a term often applied to computer directed drafting systems. Drawing devices are used in conjunction with a computer to make written or drawn presentations. See Fig. 22-1. Computer directed systems are frequently used when the volume of graphic presentations makes it impractical to perform the task manually.

The development of computer directed drawing devices has taken several directions.

COMPUTER DIRECTED TRACER SYSTEM

With the complexity of products increasing, the need for technical handbooks and manuals becomes much greater. For example, 21 different manuals are required to provide information on the operation and maintenance of one of our modern jet aircraft. The

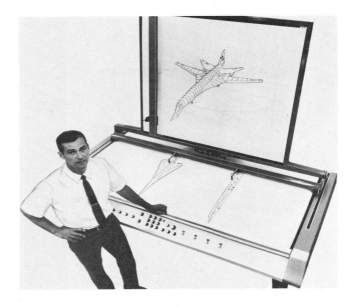

Fig. 22-2. Computer directed Illustromat 1100 is capable of making accurate perspective drawings. (Perspective Inc.)

Fig. 22-1. A typical computer center. (California Computer Products, Inc.)

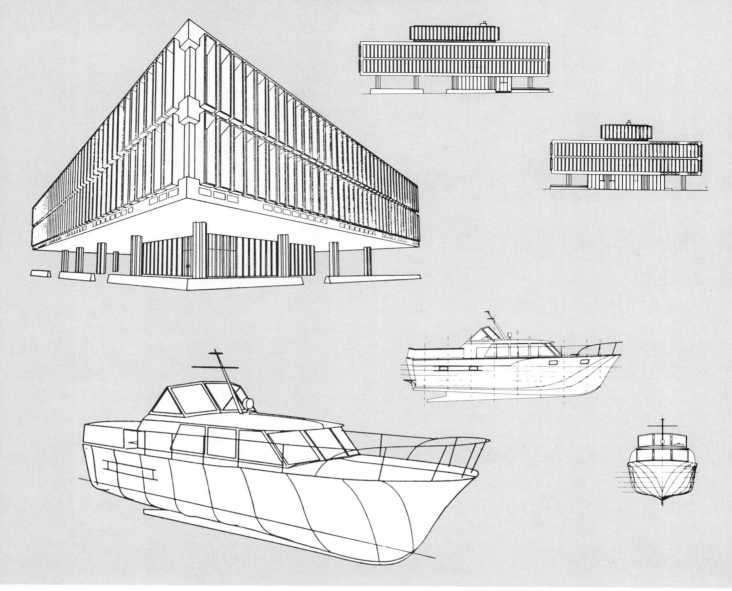

Fig. 22-3. Perspective drawings made by computer directed
instrument from two views of an engineering drawing.

manuals would make a stack seven feet high if they were piled one on another.

To meet the demand for pictorial drawings such as this, the development of automated equipment computer directed tracer systems has been necessary. The computer directed instrument shown in Fig. 22-2 is one such device. This instrument can draw mathematically and mechanically accurate perspective, isometric, and stereoscopic 3—D views.

The finished view can be given any tilt and rotation from 0 to 360 degrees. Scaled enlargements or reductions can be made. Also, with this instrument, the missing third view can be produced from a drawing having only two views.

A special preliminary drawing need not be made. The operator sets up the machine and traces two views (side and end views) from the engineering drawing. The instrument completes the perspective. See Fig. 22-3. The unit consists of three primary components, Fig. 22-4:

1. The horizontal TRACING ASSEMBLY with two tracing styluses. The styluses may be replaced by pencils.
2. The CONTROL PANEL which operates the computer.
3. The vertically-mounted PLOTTING BOARD with a motorized pen.

Machine setup is rapid and comparatively simple. Two perpendicular (blueprint) views of the subject

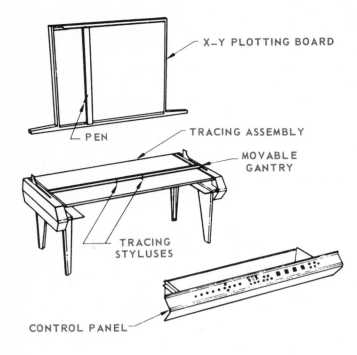

Fig. 22-4. Basic components of the Illustromat 1100. (Perspective Inc.)

are placed on the horizontal board. A sheet of paper is attached to the vertical board on which the three-dimensional view is to be drawn. The controls are adjusted to the desired station point distance, subject rotation and tilt, and the scale of the perspective drawing.

The operator traces each of the print views using the stylus. Line information from the two print views of the subject is fed into the computer. The computer converts the two-dimensional data to three-dimensional data; then directs the pen on the vertical plotting board to make the drawing. See Fig. 22-5.

Using this setup, accurate drawings (perspective, assembly, exploded view, etc.) can be completed quickly. See Fig. 22-6.

PHOTO-COMPOSING SYSTEM OF AUTOMATED DRAFTING

Another approach to automated drafting makes use of a photo-composing system, Fig. 22-7. Typical

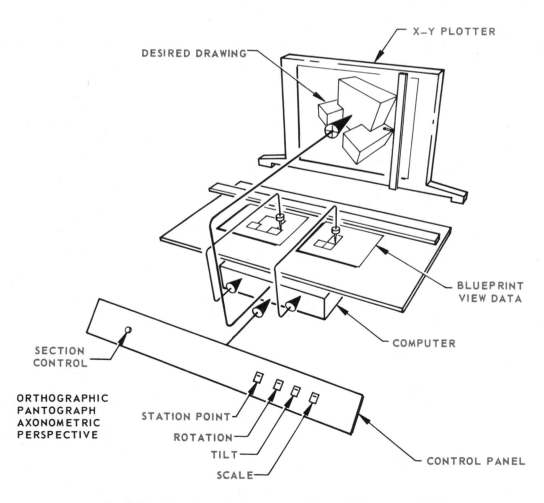

Fig. 22-5. Operation of the computer directed Illustromat 1100.

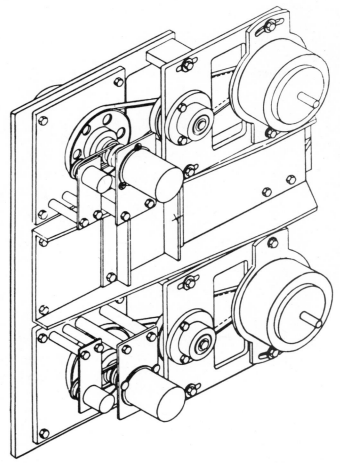

Fig. 22-6. Drawing made by computer directed device in 1.9 hours. (Perspective Inc.)

applications for this system include the repeated use of symbols such as schematics, printed circuits, flow charts, industrial piping, computer wiring, and hydraulic diagrams.

Fig. 22-7. The Mergenthaler DIAGRAMMER. (Mergenthaler Linotype Co.)

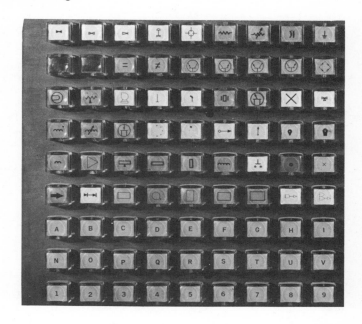

Fig. 22-8. Partial view of the 256 push buttons on the Selector Unit which causes the symbols to be projected onto the viewing screen. (Mergenthaler Linotype Co.)

To operate the machine, one of the 256 symbol buttons on the SELECTOR UNIT, Fig. 22-8, is selected by the operator and pushed. A full scale projection of the symbol is projected onto the VIEWING SCREEN. By working the controls, the image can be magnified, rotated to any angular position, and moved horizontally, vertically or diagonally over the VIEWING SCREEN.

The symbol or line (the unit can produce seven basic types of lines) is recorded on film when the operator touches the EXPOSE BUTTON. Exposed images can be maintained on the film throughout the drawing procedure for progress reference. A rough sketch is frequently used as a guide.

Lettering is placed on the film by pushing appropriate buttons. The lettering may be made to run up, down, to the right or to the left. A space bar is provided for word spacing.

When all elements have been selected, positioned and exposed, the film is removed from the operating unit for normal photographic processing. The drawing in Fig. 22-9 (reduced in size) was produced on a photo-composing unit.

COMPUTERIZED GRAPHIC DISPLAY SYSTEMS

Computer graphic display systems are among the most versatile of computer graphic techniques. They

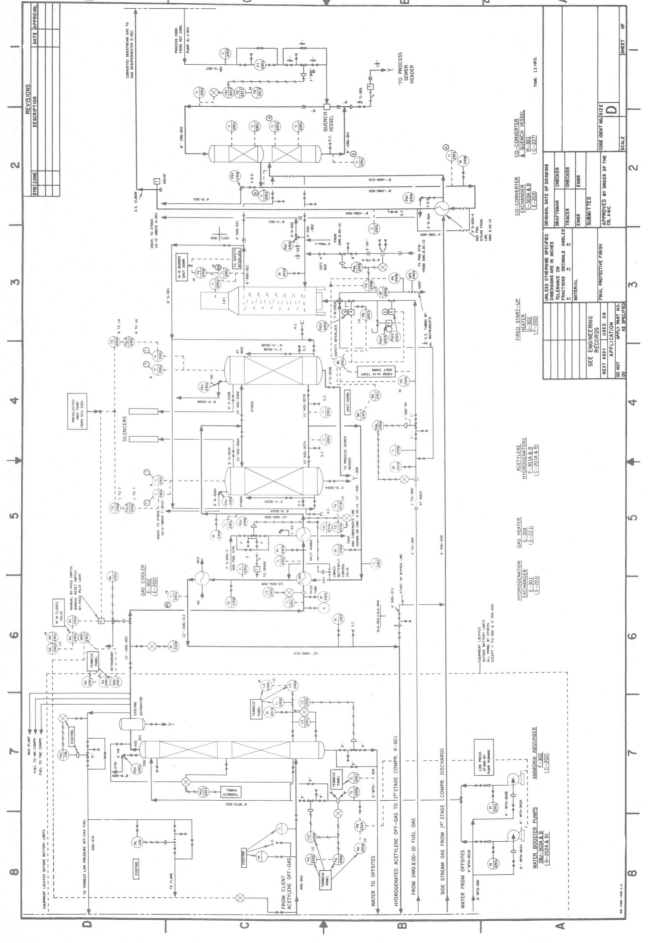

Fig. 22-9. Drawing produced by the DIAGRAMMER (greatly reduced). (Mergenthaler Linotype Co.)

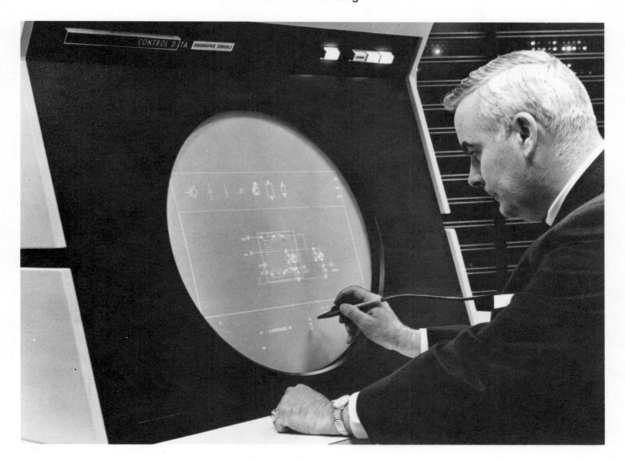

Fig. 22-10. System in which user establishes two-way communication with computer by using a light pen, alphabet letters and numbers on push buttons to present problem on cathode ray tube. (Control Data Corp.)

enable the user to establish two-way communication with the computer in alphabet letters or numbers. See Fig. 22-10.

The basic elements of such a system are:
1. The computer.
2. A "software" system (program) that provides contact between the user and the computer.
3. A CRT (cathode ray tube in which slender beams are projected onto the surface) display scope.
4. A hand-held pen, used for drawing and activating controls on the scope.
5. A keyboard with symbol push buttons.

By using the light pen and the push buttons, the problem is shown on the scope in graphic symbols — alphabet and number characters, points and lines. Curves are represented by a sequence of dots or short straight lines. The computer provides an immediate display that may be analyzed and modified if needed. Engineering changes on one drawing can be made on thousands of drawings in a computer file. For example: all hex head cap screws can be changed to socket head cap screws by a single entry.

The light pen, when pointed at any specific item (character, point or line) displayed on the screen, identifies the displayed information to the computer for proper action. Action also may be initiated by using the appropriate push button.

It is possible to write a program to enable the computer to smooth and straighten hand-drawn lines and to draw lines of requested lengths or desired angles. Any section of the diagram can be viewed at a reduced or enlarged size. Different views of the same figure can be viewed at one time, and a geometric figure can be rotated to any desired viewing angle.

The system can be adapted to virtually any situation that involves calculation and manipulation of data or symbols.

NUMERICALLY CONTROLLED MACHINE TOOLS

Preparing tapes which operate numerically controlled machines is another important application of computer graphics.

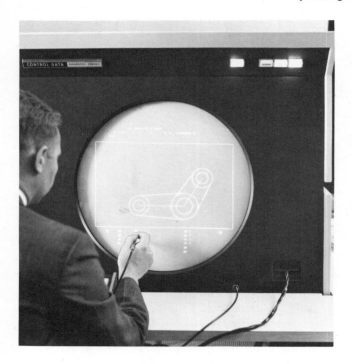

Fig. 22-11. Hand-held light pen is used to trace path of tool cutter around part. This activates computer which automatically punches out tape that controls operation of machine tool.

Fig. 22-12. An electrical field in the "glass sandwich" that makes up the writing surface of the Data Tablet senses the location of the stylus 200 times a second. Depending on the electronics involved, the movement of the pen is converted into signals which perform a wide range of functions. (Sylvania Electric Products Inc.)

In using the system shown in Fig. 22-11, a description of the part to be produced is entered on the cathode ray tube from a print of the finished part. The path of the machine cutting tool is traced around the part, using a hand-held light pen. The computer then punches out a control tape that will enable the machine to make the required cuts.

SYLVANIA DATA TABLET

The SYLVANIA DATA TABLET[1], Fig. 22-12, is a unique computer directed system. With it, data processing is joined with basic pen-and-pencil methods of processing and transmitting information. The unit enables the computer and other electronic equipment to accept handwritten information. It can add to or replace conventional input techniques (punched cards and paper or magnetic tapes).

The Data Tablet system consists of an electronic stylus and a writing surface composed of a transparent conductive film sandwiched between protective layers of glass. See Fig. 22-12. Direct current voltages proportional to position are impressed on the conductive layer so that coordinate position ("X" — horizontal axis, "Y" — vertical axis) can be

sensed by the stylus as it moves over the tablet surface. The position of the stylus is recorded or sampled 200 times a second providing a precision of image. Accuracy of line recording is less than one percent deviation from the actual drawing.

A small cabinet contains the associated solid state electronics that permits the unit to be used with almost any computer.

Fig. 22-13. The Data Tablet can be used at a desk, console or as a transparent overlay for a CRT. (GTE Sylvania)

[1]Developed by the Applied Research Laboratory of GTE Sylvania, Waltham, Massachusetts.

Since the writing surface is transparent, the tablet can be placed over the image on the CRT and used, for example, for designating positions on a military map. See Fig. 22-13. This usage would permit combat commanders to be kept up to date on the latest developments on the battlefield.

The unit can also be used in the field of publishing. Copy readers can display portions of a manuscript and make alterations and corrections with the stylus alone or with keyboard entries.

An important feature of the Data Tablet is its ability to simplify and expedite the flow of information into computers. It is possible to feed information directly into a computer, ask questions, obtain drawings, and make revisions merely by drawing symbols or schematics. This eliminates the need for having to write new computer programs and prepare new punch cards or tapes.

GLOSSARY OF COMPUTER GRAPHIC TERMS

ADDRESS: A number, name or label identifying a specific location within the computer's memory apparatus.

ALPHANUMERIC: Pertaining to a set of characters that contains both letters and numbers.

ANALOG: Denotes the use of physical variables (distance, rotation or voltage) to represent and correspond with numerical variables that occur in computation.

ANALOG COMPUTER: A computer that operates on analog data by performing physical processes on these data.

CATHODE RAY TUBE (CRT): A vacuum tube in which cathode rays are used to produce luminous spots on its surface. Similar to the picture tube in a TV set.

COMPUTER: A device capable of solving porblems by accepting data, performing prescribed operations on the data, and supplying the results of these operations.

CONSOLE: Location of computer controls, as well as various lights and the cathode ray tube display.

COORDINATES: The positions or locations of points or planes.

DATA: Facts or information taken in, acted upon or emitted by a machine used for handling information.

DATA TABLET: A unique computer directed system that accepts handwritten information in place of — or in addition to — conventional input techniques.

DIGITAL COMPUTER: A computer which produces results from numeric information only, and performs operations by means of counting, rather than measuring as in analog computers.

GRAPHIC: Written or drawn.

HARDWARE: The computer and its accessories.

KEYBOARD: Part of a device that punches holes in a card or tape to represent data, or a device that communicates directly with a computer.

LIGHT PEN: A hand-held, pen-like device containing a photocell or photo multiplier, used for the generation of lines on a display.

MACHINE LANGUAGE: Instructions written as binary (base two) codes.

MEMORY: A term referring to the equipment and media used for storing information (data and instructions) in machine language in electrical or magnetic form.

NUMERICAL CONTROL SYSTEM: A system in which actions are controlled by the direct insertion of numerical data.

OPTICAL SCANNER: A device that optically scans (examines) printed or written data and generates electrical representation for input to the computer.

PHOTO-COMPOSING SYSTEM: With this system, the operator selects data to be projected onto a viewing screen, then positions and exposes the data to record them on film. The film is removed from the operating unit for normal photographic processing.

PLOTTER: Automatic drawing equipment controlled by a tape or directly by a computer.

PROGRAM: A set of instructions for the computer that defines a desired sequence of conditions for a process or function, and the operations required between these conditions.

PUNCHED CARDS: Cards of uniform size and shape, suitable for punching in a meaningful pattern and for mechanical handling. The punched holes are usually sensed electrically or mechanically.

PUNCHED TAPE: Paper or plastic tape into which a pattern of holes has been punched to convey information.

REAL TIME: The time the computer needs to respond with a solution.

SOFTWARE: The means of communicating with the machine. The program.

STYLUS: An instrument used to record lines on the writing surface of a Sylvania Data Tablet or computer directed tracing system.

TIME-SHARING: Process in which the computer switches rapidly from one problem to another, giving a number of human users the illusion of working on each individual problem all of the time.

TEST YOUR KNOWLEDGE - UNIT 22

(Write answers on a separate sheet of paper.)

1. In computer graphics (automated drafting) drawing devices are used in connection with a computer to make _____ and _____ presentations.

2. Computer directed instruments can draw mathematically and mechanically accurate perspective, isometric and stereoscopic 3D views. True or False?

3. In the photo composing system of automated drafting, typical applications include the repeated use of _____.

4. In a system in which the user establishes a two-way communication with a computer, the operator must use a _____, _____ and _____ with push buttons.

5. In preparing tapes which operate numerically controlled machines, a description of the part to be produced is entered on the _____ from a print of the finished part.

6. A cathode ray tube is a vacuum tube in which cathode rays are used to produce _____ on its surface.

Industry photo. The architect who designed this building communicates, through his drafting ability, with these carpenters by means of blueprints.

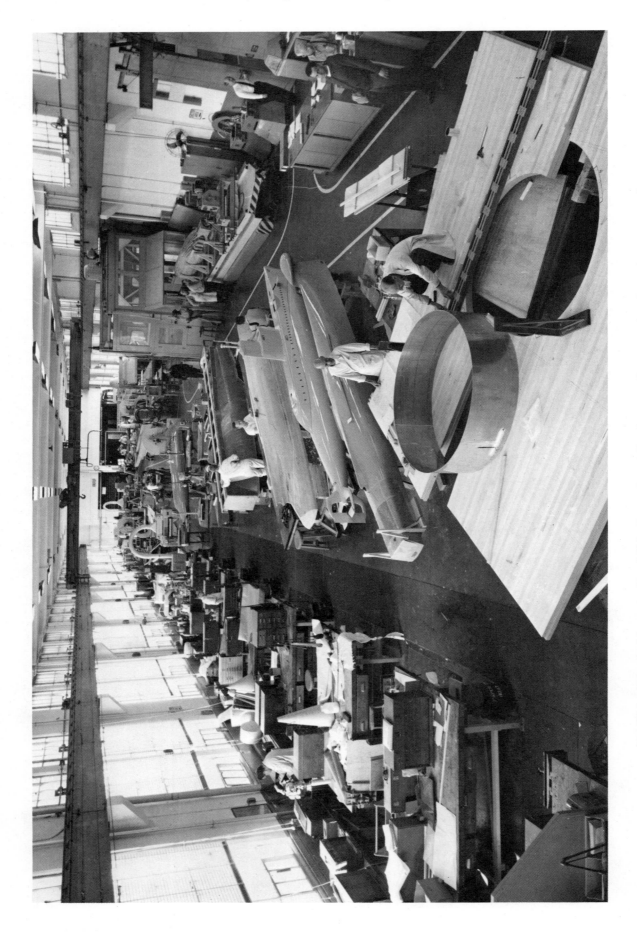

Fig. 23-1. Ship models at the Naval Ship Research and Development Center used to determine a proposed ship's seaworthiness, power requirements, resistance to water, and stability characteristics.

Unit 23
MODELS, MOCKUPS, AND PROTOTYPES

To many people, modelmaking is an interesting hobby. You probably have made models of famous planes, boats or cars from wood or plastic.

Industry makes extensive use of three model-making activities: MODELS, MOCKUPS and PROTO-TYPES. These are used for engineering, educational and planning purposes. Models have proved to be very helpful in solving design problems and to check the workability of a design or idea before it is put into production.

MODEL. Industry's definition of a "scaled model" is a replica of a proposed, planned or existing object. See Fig. 23-1. The model may be constructed to see how the product will look, to check out scientific theory, demonstrate ideas or for training or advertising purposes.

MOCKUP. A full-size, three-dimensional copy of an object, Fig. 23-2. The mockup usually is made of plywood, plaster, clay, fiber glass, plaster or a combination of materials.

Fig. 23-2. Mockup of the ASTRO III, a turbine powered, three-wheel "dream" car. Ideas from this car may be used in future auto production.   (Chevrolet)

Fig. 23-3. The TURBOTRAIN is the prototype of trains that will furnish high speed transportation. It is powered by a gas turbine and is capable of reaching speeds up to 170 mph. (Sikorsky Aircraft)

PROTOTYPE. A full-size operating model of the production item, Fig. 23-3. It usually is handcrafted to check out and eliminate possible design and production "bugs."

HOW MODELS, MOCKUPS AND PROTOTYPES ARE USED

Many industries employ models, mockups and prototypes for design tools. See Fig. 23-4. A few of the more important applications are: automotive, aerospace, architecture, ship building, city planning and construction engineering.

AUTOMOTIVE INDUSTRY

The automotive industry places great importance on the use of models, mockups and prototypes. Mistakes can be very costly.

Fig. 23-4. An automobile design studio is a busy place, with various members of the design team working simultaneously on different stages of the design process. In the foreground, a designer works out a detail idea in sketches. In the background, the shape of an experimental GM—X is taking form in a full-size clay model at the hands of modelmakers, working under the direction of the chief designer. (General Motors)

Fig. 23-5. The automotive industry uses sketches to develop ideas. Scale models are made of promising ideas for further study. (Oldsmobile)

An automobile starts "life" as a series of sketches developed around specifications supplied by management. Occasionally, the more promising sketches are drawn full scale for additional evaluation.

Clay models are used for three-dimensional studies. Upon the completion of further design development, a fiber glass prototype usually is constructed, Fig. 23-6.

Fig. 23-6. A fiber glass prototype of an automobile that will feature a plastic body. (Marbon Div., Borg-Warner)

261

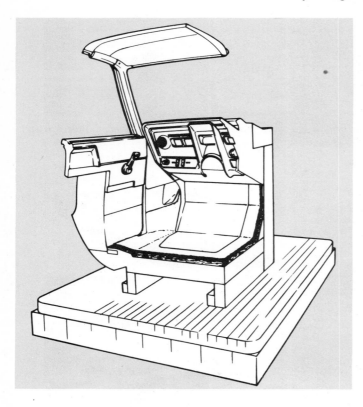

Fig. 23-7. Wood models (called "bucks") are used to prove the accuracy of molding and trim designs. (General Motors)

Fig. 23-9. A few of the many shuttle re-entry vehicle design ideas tested in model form. To test full size aircraft would have cost millions of dollars and endangered the lives of many people. (NASA)

Production fixtures (devices to hold body panels and other parts while they are welded together) and other tools needed are developed from accurate, full size wood and plaster models. See Fig. 23-7.

After many months of development and production planning, manufacturers are ready to make new model automobiles available to the public.

AEROSPACE APPLICATIONS

Cost makes it mandatory that aerospace vehicles first be developed in model or mockup form, Fig. 23-8.

Flight characteristics can be determined with considerable accuracy, without endangering human life, by testing complicated models in a wind tunnel or in free flight, Fig. 23-9.

Fig. 23-8. A full-scale mockup of the supersonic Concorde jet airplane. Using mockups such as this, ideas can be checked without the great expense of reworking the actual plane. (Trans World Airlines)

Fig. 23-10. Models were used when planning moon landings.

Exploration of the moon was first planned with models. See Fig. 23-10.

Prototype aircraft, Fig. 23-11, are the first two or three preproduction planes (usually handcrafted) that are flown to check the data obtained from wind tunnel research and to secure a license for that particular type plane.

ARCHITECTURE

You have seen photos of proposed buildings in the real estate section of your Sunday paper. Some of these illustrations were of models that were very accurate minature replicas of the proposed buildings. See Fig. 23-12.

Many people use models when planning a new home, Fig. 23-13. This helps them to visualize how the house will look when completed. The model enables the owner to see the completed design in three dimensions and also how paint colors and shrubbery plantings will look.

SHIP BUILDING

Ship hulls are tested in model form, Fig. 23-14, before designs are finalized and construction starts. Specially designed equipment tows the model hull through the water in the test basin. The model hull behaves like the full-size ship, so design faults can be located and corrected.

CITY PLANNING

Most large cities use scale models to show city officials and planners how proposed changes and future developments will look. Models, while rather costly, permit intelligent decisions to be made before large sums of money are spent acquiring land and existing buildings are torn down.

CONSTRUCTION ENGINEERING

Many construction projects are designed from carefully constructed models, Fig. 23-15. By working from models, engineers can see how space can best be utilized. In some instances, they can determine how

Fig. 23-11. Prototype of the 747 was used to flight prove design data.   (Boeing)

Fig. 23-12. Architectural model of the Bel Air, Maryland, high school (where the author teaches).

Fig. 23-13. Many potential home builders make models such as this one. They can see how their home will look before starting to build. (Taylor Made Models)

Fig. 23-14. Left. Model of a U.S. Coast Guard vessel in seaworthiness test in the basin at the Naval Ship Research and Development Center. (U.S. Navy) Fig. 23-15. Right. Highly detailed model of a power generating plant. Note the clear plastic walls. The model was used to train operating personnel.

the proposed project will affect the surrounding community. See Fig. 23-16. This helps to minimize field problems and changes during construction.

Models also may be used to train personnel in plant operation.

CONSTRUCTION MODELS

Model making materials are readily available commercially, Fig. 23-17. Many products made for the model railroader are ideally suited for making architectural models. Kits are available for the small home builder who wants to design his own home.

A professional touch can be added to models by using accurately scaled furniture, automobiles and figures that can be purchased at toy and hobby shops.

Preprinted sheets of brick and stone can be glued to a suitable thickness of balsa wood for walls and partitions. Various types of abrasive paper are suitable for roofing, driveways and walkways. Simulated window glass can be made from transparent plastic

Fig. 23-17. A few of the model making materials available at hobby shops.

sheet. Several different scale sizes of window and door frames are available molded in plastic. Most model shops can supply bushes and trees in various types and sizes.

Other types of models — autos, planes and boats — are made from bass wood, mahogany, balsa wood, metal, plaster and various kinds of plastics.

Regular model making paint is produced in hundreds of colors and is ideal for painting all types of models. However, care must be exercised when painting models that have plastic in their construction. Be sure the paints used are designed for plastics. If you are not sure, paint a small portion of the plastic that is hidden from view to determine whether the paint is compatible with the material.

Models may be assembled with model airplane cement or the white glues (Elmer's, Titebond, Ross, etc.).

TEST YOUR KNOWLEDGE - UNIT 23

(Write answers on a separate sheet of paper.)

1. Industry uses models, mockups and prototypes for what purposes?
2. A MODEL is _____.
3. A MOCKUP is _____.
4. A PROTOTYPE is _____.
5. List four industries that employ models, mockups and prototypes as design tools. Briefly describe how each industry uses them.

Fig. 23-16. Model of a portion of the Susquehanna River in Maryland to test how the proposed atomic power generating plant will affect the river and its ecology (branch of biology dealing with relations between animal and plant life and their environment).   (Philadelphia Electric Co./Alden Research Lab)

Fig. 23-18. Plastic model helicopter to be used as a basis for constructing a hanger.

6. Walls and partitions in a model home usually are constructed of _____ wood.
7. Roofs, driveways and walkways can be made from different kinds and grades of _____.
8. Models of cars, planes and boats may be made from what materials?

OUTSIDE ACTIVITIES

1. Visit a model or hobby shop, then prepare a list of available materials which may be used for building architectural models.
2. Review technical magazines and clip illustrations that show models, mockups and prototypes being used for engineering, educational, planning or other purposes. (Do not cut up library copies.)

3. Make a collection of materials suitable for building model homes.
4. Visit a professional model maker in your community. With permission, make a series of slides showing examples of various models and how these models are made. Then, give a talk to your class on professional model making.
5. Make a scale model of your drafting room. Discuss alternate layouts.
6. Construct a model helicopter similar to the one shown in Fig. 23-18. Design and construct from balsa wood a minimum building that would protect the helicopter from the elements.
7. Design and construct a full-size model of a disposable waste basket, using corrugated cardboard.

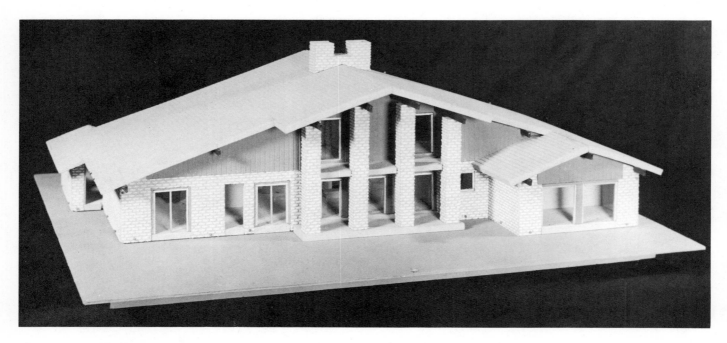

Home ideas can be seen in miniature with a model. Costly changes can be eliminated before construction begins. (Taylor-Made Models)

Model showing the details of the tip of Manhattan and a portion of Brooklyn. This model will be used for planning purposes. Many drafters will be needed to plan the future expansion of the city. (Lester Associates, Inc.)

Unit 24
DRAFTING AIDS

DRAFTING AIDS are devices and techniques that help drafters do their work better and/or in less time. Typical aids commonly used by industry are included.

TEMPLATES

TEMPLATES, Fig. 24-1, are probably the most widely used of the drafting aids. These timesaving drafting tools are available in almost an unlimited range of standard symbols and figures. See Fig. 24-2. The templates are made of transparent plastic.

GRIDS FOR GRAPHS AND SCALE DRAWINGS

A graph is a picture of a collection of facts. Many graphs are plotted on the familiar square or cross-sectioned grid paper (See Unit 16).

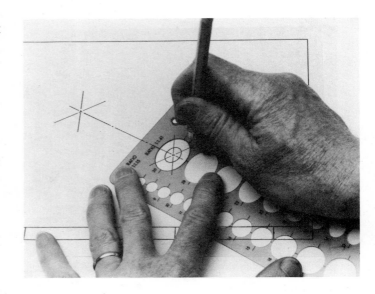

Fig. 24-1. Templates enable the drafter to do normally time-consuming jobs with ease and accuracy.

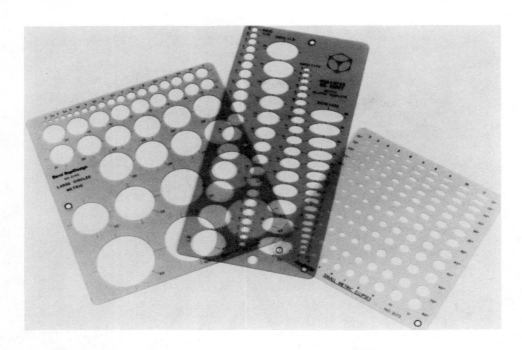

Fig. 24-2. A few of the many different types of templates used in the modern drafting room.

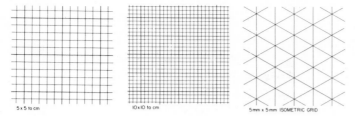

Fig. 24-3. Graph paper is manufactured in many different cross-sectioned grid sizes.

In many instances, drawing the grid (cross-sectioned background) is more time consuming than plotting and drawing in the graph or chart. Many types and sizes of grid patterns are available commercially in the form of prepunched 8 1/2 in. by 11 in. loose leaf sheets and in other dimensions. See Fig. 24-3. The grid lines usually are printed in green ink. However, some grid patterns are printed in orange, pale blue or black.

Designers, drafters, surveyors, engineers and architects also make considerable use of commercially prepared graph sheets to make preliminary design studies. See Fig. 24-4. For example, the isometric grid makes it very easy to convert an orthographic drawing into an isometric drawing, Fig. 24-5.

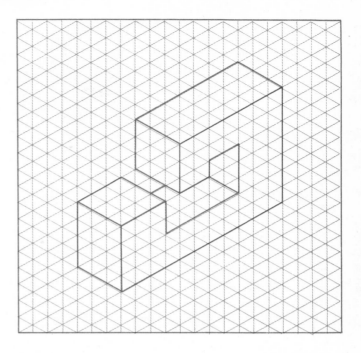

Fig. 24-5. An isometric drawing made on graph paper designed for that purpose. A 5 mm x 5 mm grid was used.

PREPRINTED MATERIALS

Preprinted materials are another means of simplifying work and saving time for the drafter. Drafting

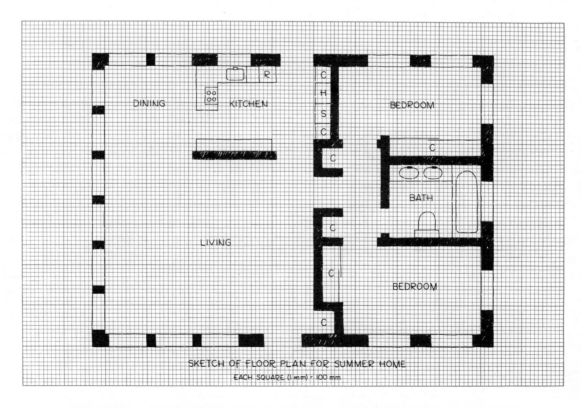

Fig. 24-4. A floor plan sketched on graph paper. Changes can be made easily on the grid. Each square (1.0 mm) equals 100 mm.

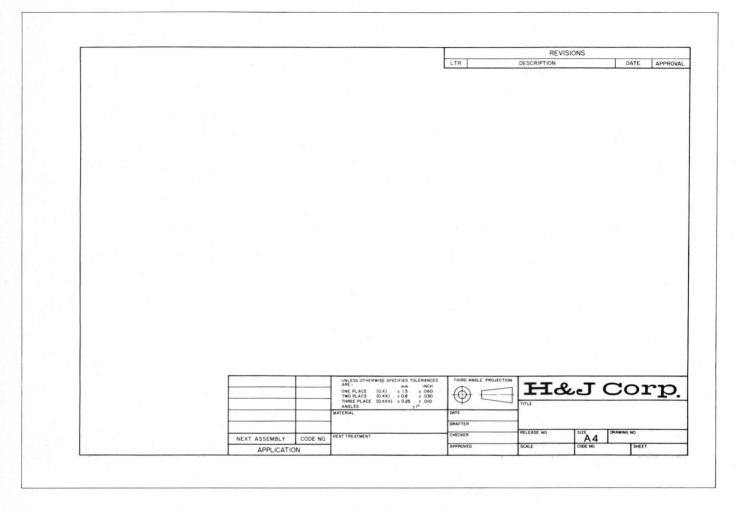

Fig. 24-6. Preprinted drawing sheets save the drafter's time. All sheets are identical in format.

aids and special materials that fall into this category are available in the following forms:

1. PREPRINTED DRAWING SHEETS, Fig. 24-6, are imprinted drawing forms that come in standard sheet sizes and most types of drawing media.
2. TITLE BLOCKS preprinted on acetate or Mylar sheet are widely used, Fig. 24-7. One side of the plastic sheet is coated with a pressure sensitive adhesive, making it easy to apply to the drawing sheet. Title blocks are printed to order.
3. GRAPHIC SYMBOLS, Fig. 24-8, are used to simplify repetitive work. These are applied to the drawing sheet by means of pressure sensitive adhesive or by the transfer method. The symbol sheet is placed in position on the drawing and the symbol is transferred by burnishing (rubbing) the backing sheet with a smooth burnishing tool, Fig. 24-9. The transferred symbol may be removed with a soft pencil eraser.

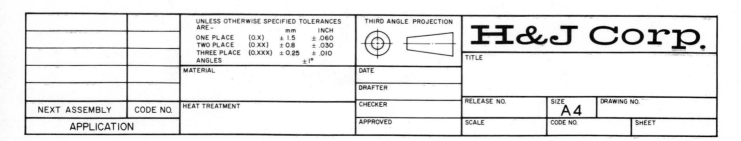

Fig. 24-7. Preprinted title block.

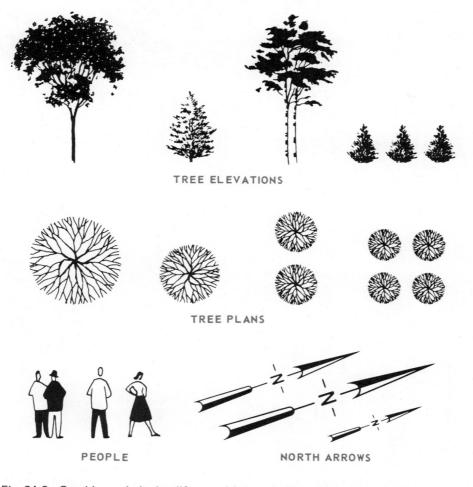

TREE ELEVATIONS

TREE PLANS

PEOPLE   NORTH ARROWS

Fig. 24-8. Graphic symbols simplify repetitive work. Many designs are available.

Fig. 24-9. Applying a graphic symbol to an architectural drawing by burnishing (rubbing) it into position.

4. Preprinted LETTERING which is available in hundreds of styles and sizes, enables students to do professional type lettering, Fig. 24-10. The material is manufactured in TRANSFER LETTERING and in CUT-OUT LETTERING. With transfer lettering, the letters are transferred to the drawing by burnishing the back of the sheet holding the letters, Fig. 24-11. With cut-out lettering, the individual letters are cut out and held in place by pressure sensitive adhesive, Fig. 24-12.

5. SHADING MEDIUMS, Fig. 24-13, make it possible to show differences in areas in mapping, Fig. 24-14, and to highlight areas. There are patterns for indicating swamps, forests and other areas.

6. PRESSURE SENSITIVE TAPES, Fig. 24-15, supplement conventional drafting instruments, for making charts, graphs, electronic diagrams, etc. Many different colors, styles and widths of pressure sensitive letters and symbols are available for use on drawings. The tapes save time and effort.

LEADERSHIP IN THE
18 leadership in the creat
LEADERSHIP IN
24 leadership in the

LEADERSHIP
24 leadership in
LEADERS
30 leadership

Leaders
Leadershi
leadership in
Leadership In T

LEADERSHIP IN THE
24 leadership in the cre
LEADERSHIP IN
36 leadership in
LEADERSHIP IN THE
24 leadership in the
LEADERSHIP

Fig. 24-10. Preprinted lettering. Many styles and sizes are produced.   (Formatt)

Fig. 24-11. Left. Transfer letters are applied by placing the letter in position and burnishing the back of the sheet with a smooth object. Right. Removing the lettering sheet after the letter has been applied to the sheet. The guide line is erased after the entire line of lettering has been applied.

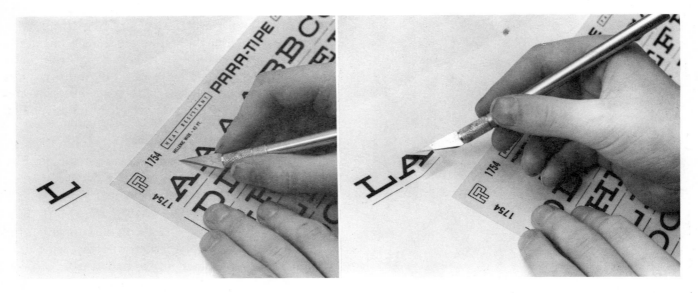

Fig. 24-12. Left. Removing a CUT-OUT LETTER from the lettering sheet. Right. Applying the letter to the sheet. The guide line is cut away after the entire line of lettering has been set into position.

Fig. 24-13. Many patterns of shading mediums are available.

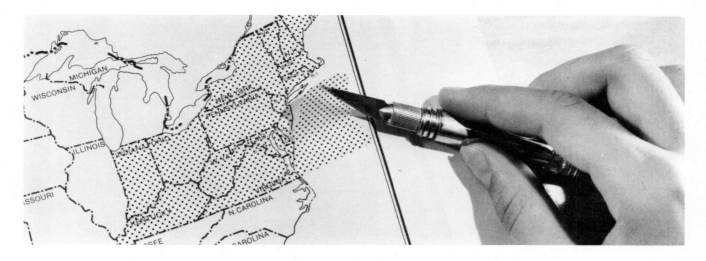

Fig. 24-14. Shading media are used to differentiate between areas on a map.   (Formatt)

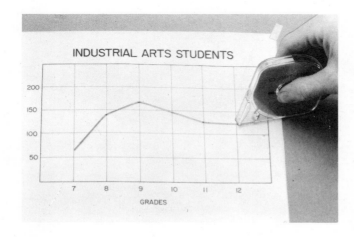

Fig. 24-15. Graphs are easy to make by using pressure sensitive tapes. Different color tapes may be used to provide different bits of information. They are available in many styles and in a variety of widths.

TEST YOUR KNOWLEDGE - UNIT 24

(Write answers on a separate sheet of paper.)

1. Drafting aids are _____.
2. Templates are:
    a. Convenient and timesaving drafting tools.
    b. Available in a great number of standard styles and symbols.
    c. Tools which enable the drafter to do normally time-consuming jobs with ease and accuracy.
    d. All of the above.
3. Match these words with the letters of the correct descriptions that follow:
    _____Grids.
    _____Preprinted title blocks.
    _____Preprinted drawing sheets.
    _____Graphic symbols.

_____Preprinted lettering.
_____Shading mediums.
_____Pressure sensitive tape.

A. Used to simplify repetitive work.
B. Supplement conventional drafting tools.
C. Enables students to do professional type lettering in various styles and sizes.
D. Preprinted on Mylar or acetate sheet.
E. Available in all standard drawing sheet sizes.
F. Used to show difference in areas in mapping, or to highlight areas on a drawing.
G. Often used to plot charts and graphs.
H. Made of transparent plastic and are available in an almost unlimited range of standard symbols and figures.

OUTSIDE ACTIVITIES

1. Secure samples of the following:
   a. Graph paper.
   b. Pressure sensitive tapes.
   c. Preprinted symbols.
   d. Preprinted lettering.
   e. Preprinted title blocks.
2. Get samples of work done on a preprinted drawing sheet.
3. Prepare one of the graphing problems in Unit 16, using pressure sensitive tapes.
4. Contact manufacturers of pressure sensitive drafting aids for catalogs to be kept in your drafting room library.
5. Demonstrate the proper use of templates.

Unit 25
DESIGN IN INDUSTRIAL ARTS

In many ways, good design may be considered a carefully thought out plan for a direct solution to a problem. This is sometimes called CREATIVE PLANNING. The development of the plan includes creating, inventing and research. Usually many ideas are studied, tried-out, analyzed and then either incorporated into the design or discarded.

DESIGN GUIDELINES

Though design is not an exact science, good design is characterized by certain qualities that are easily recognized:

FUNCTION. How well does the design fit the purpose for which it was planned? Does it fulfill a need? Function often dictates the product's appearance, Fig. 25-1.

HONESTY. Is there an honest use of material? The qualities of the material (strength, texture, etc.) should be emphasized to the fullest. The product should be able to do the job it was designed to do, Fig. 25-2.

APPEARANCE. Do the individual parts of the design create interest when they are brought together? See Fig. 25-3. Are the proportions in balance and do the components belong together? Is the product pleasing in appearance?

Fig. 25-2. There is an honest use of materials in this speed boat. The fiber glass hull is light, of high strength, and it is not affected by salt water. Color is molded in the hull, and it is easy to clean. (Evinrude Motors)

Fig. 25-1. The job for which this all-terrain TERRA TIGER was designed (for operation on rough ground, water and snow) pretty well dictated how it should look. (Allis Chalmers)

Fig. 25-3. This copy of a colonial American water pitcher is a fine example of good design. Its proportions are in balance, its contours are smooth and it is pleasing in appearance.
(Shirley Pewter Shop, Williamsburg)

QUALITY OF FINISHED WORK. The care with which the project is put together is an inherent part of good design. Quality must be built into the work, not added on.

Simply stated, good design is distinguished by certain recognizable qualities. All of the qualities are necessary. Overlooking or eliminating one may destroy the entire design.

DESIGNING A PROJECT

How should YOU go about designing a project? The easiest approach is to follow a pattern similar to that used by a professional designer. Think of the project as a DESIGN PROBLEM:
1. STATE THE PROBLEM. What is the purpose of the project? How will it be used?
2. THINK THROUGH THE PROBLEM. What must the project do? How can it be done? What are the limitations that must be considered? How have others solved similar problems? Study comparable products for ideas, Fig. 25-4.
3. DEVELOP YOUR IDEAS. Next, make sketches of your ideas, Fig. 25-5. Develop the ideas and have them criticized by your teacher and fellow stu-

Fig. 25-4. This young man wants to design a particular kind of model airplane. He is studying examples of models that others have designed and built. He is thinking through the design problem.

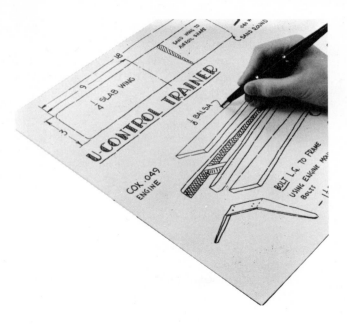

Fig. 25-5. Making sketches of ideas.

dents. Determine what materials will be best suited for the project.

4. MAKE MODELS. Put your best ideas into model form. Develop fabrication techniques for your ideas.

5. PREPARE WORKING DRAWINGS. After you are satisfied that you have solved the problem to the best of your ability, prepare working drawings for the project.

6. CONSTRUCT THE PROJECT to the best of your ability. Do a job that you will be proud to show to others. See Fig. 25-6.

The development of a well-designed product (project) takes time. Do not be discouraged if your first attempts at designing fall short of your goal. It takes time to acquire the skill and ability to solve design problems. You are urged to read and study the many fine publications on design to acquaint yourself with numerous well-designed products that are available.

Keep a notebook of your ideas and ideas you have clipped from newspapers, advertising folders and magazines (but not those in the library). Include photos of projects you have designed and constructed to make it easier to evaluate your work and to help you see your improvement.

TEST YOUR KNOWLEDGE - UNIT 25

(Write answers on a separate sheet of paper.)

1. A well-designed product should serve several purposes:
   a. _____.
   b. _____.
   c. _____.
2. Good design is hard to describe. Why?
3. Good design is characterized by certain qualities

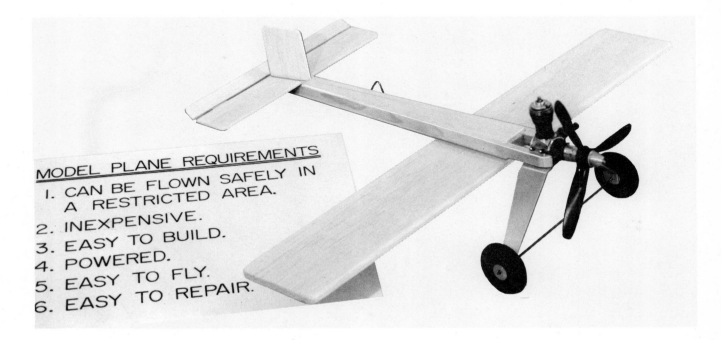

Fig. 25-6. The completed model ready for painting, trimming and test flying. The model meets design requirements.

or guidelines. Describe the guidelines briefly.

4. When solving a design problem, a series of steps should be followed. List and briefly describe the steps.

OUTSIDE ACTIVITIES

1. The design problems listed below are presented to get you started. You may select one and develop it, or you may originate a problem of your own.

   a. Shoeshine box.
   b. Coffee table.
   c. Table lamp.
   d. Hand launched glider.
   e. Jet propelled (CO_2 cartridge or Jetex Rocket engine) model automobile.
   f. Disposable waste basket (made from corrugated cardboard).
   g. Clock (movement may be purchased from School Products Co., 1201 Broadway, New York, NY 10001).
   h. Book rack.
   i. Tray.
   j. Wall shelf.
   k. Salad server (laminated wood, wood, plastic, metal).
   l. Tool box.
   m. Turned lamp.
   n. Turned bowl.
   o. Plastic soft food spreader.
   p. Bird house.
   q. Fiber glass model boat.
   r. Cutting board.
   s. Metal lamp.

Unit 26

DRAFTING AND DESIGN CAREERS

If you like drawing and designing as described in this book, why not consider a job in one of the drafting and design careers?

There is a great need for drawings. For example, more than 27,000 drawings are needed to manufacture an automobile. The field of drafting is providing employment for over one million men and women. The work of other millions requires them to be able to read and interpret drawings. See Fig. 26-1.

DRAFTING CAREERS

DRAFTERS make working plans and detailed drawings. They prepare these drawings from specifications and information received verbally and from sketches and notes.

The drafter usually starts out as a TRAINEE DRAFTER, redrawing and repairing damaged drawings. The trainee may revise engineering drawings or

Fig. 26-1. Aircraft builders must be able to read and understand drawings.
(Cessna Aircraft Corp.)

Fig. 26-2. An INDUSTRIAL DESIGNER prepares preliminary sketches of an experimental car. (Oldsmobile)

make simple detail drawings under the direct supervision of a senior drafter.

During the training period, the trainee drafter may be enrolled in a formal shop program and take classroom courses within the company or at a community college or technical school.

Upon completion of the training period, the trainee drafter advances to JUNIOR DRAFTER. Similar positions are known as: DETAILER, DETAIL DRAFTER and ASSISTANT DRAFTER.

This position calls for the preparation of detailed and working drawings of machine parts, electronic/electrical devices, structures, etc., from rough design drawings. It also may require preparation of simple assembly drawings, charts and graphs, including simple calculations made according to established drafting room procedures.

From junior drafter, the next step forward is DRAFTER. The drafter applies independent judgment in the preparation of original layouts with intricate details. A drafter must have an understanding of machine shop practices, the proper use of materials and be able to make extensive use of reference books and handbooks.

With experience, the drafter will become a SENIOR DRAFTER and do complex original work.

In time, the senior drafter may become LEAD or CHIEF DRAFTER and be responsible for all work done by the department.

Most drafters specialize in a particular field of technical drawing — aerospace, architecture, structural, etc. Regardless of the field of specialization, drafters must be able to draw rapidly, with accuracy and neatness. They also must have a thorough understanding of the materials and manufacturing methods found in their field of specialization. In addition, drafters must have a working knowledge of mathematics, science, English, materials and manufacturing processes.

The manufacturing industries employ large numbers of drafters. Others are employed by architectural and engineering firms and our government.

INDUSTRIAL DESIGNERS

The work of the INDUSTRIAL DESIGNER has considerable influence on virtually every item used in our daily living, whether it is the design of a small tape player or of a giant jet plane. Many top designers are on the staffs of major automobile manufacturers, Fig. 26-2.

In general, the chief function of the industrial designer is to simplify and improve the operation and appearance of industrial products. Design simplification usually means fewer parts to wear or malfunction. Appearance plays an important role in the sale of a product. The industrial designer must be aware of changing customs and tastes; must know why people buy and use different products.

It is recommended that prospective designers have an engineering degree and a working knowledge of engineering and manufacturing techniques and materials and their properties. They must be a good drafter and have artistic ability.

TOOL DESIGNERS

The tool designer originates the designs for cutting tools, special holding devices (fixtures), jigs, dies and machine attachments that are needed to manufacture industrial products.

Tool designers also may be responsible for making the necessary drawings, or supervise others in making them. Therefore, they must be familiar with machine shop practices, be accomplished drafters and have a working knowledge of algebra, geometry and trigonometry.

Fig. 26-3. Teaching is a challenging profession that offers considerable freedom. (Harford County, Maryland, Vo/Tech High School)

As industrial technology expands and more automated machinery is introduced, there will be a constantly increasing demand for competent tool designers.

TEACHERS

TEACHING, Fig. 26-3, is a satisfying profession which is often overlooked by students. Teachers of industrial arts, vocational and technical education are in a fortunate position. They work in a challenging profession that offers a freedom not found in many other professions.

Four years of college training are needed. Also, while industrial experience is ordinarily not required, it is highly recommended.

ENGINEERS

The ENGINEER usually specializes in one of the many branches of the profession. There are at least 25 engineering specialties recognized: aeronautical, industrial, chemical, structural, civil, electrical and metallurgy to name but a few. See Fig. 26-4.

The engineer provides the technical and, in many instances, the managerial leadership in industry government. Depending upon area of specialization, the

Fig. 26-4. Engineers tend to specialize. The structural engineer is responsible for the design of structures such as this bridge in Philadelphia. (Bethlehem Steel Co.)

281

engineer may be responsible for the design and development of new products and processes, plan structures and highways, or work out new ways of transforming raw materials into saleable products.

Laws in all 50 states and the District of Columbia provide for the licensing of engineers whose work may affect life, health and property.

A professional engineering license usually requires graduation from an approved engineering college, four years of experience and passing an examination. Some states will accept experience in place of a college degree.

ARCHITECTS

In general, ARCHITECTS plan and design all kinds of structures. However, they may specialize in specific fields of architecture: private homes, industrial buildings, schools and dormitories, commercial buildings, institutions, banks, churches, etc.

When planning a structure, the architect first consults the client on the purpose of the building, its size, location, cost range and other requirements. Upon completion and approval of preliminary drawings, detailed working drawings and specification sheets are prepared. As construction progresses, the architect usually makes periodic inspections to determine whether the plans are being followed and construction details are to specifications.

Architects must be licensed in most states. This requires graduation from an approved college program, several years of experience (similar to the internship of a physician) and the passing of a special examination.

Several years of experience are acceptable in a few states in place of graduation from an architectural program.

MODELMAKERS

Industry makes extensive use of models, mockups and prototypes for engineering, educational and planning purposes. Preparation of the models is often the responsibility of the engineering drafting group, although professional MODELMAKERS are frequently used.

The modelmaker makes scale models of the proposed building or product to show the client what it will look like and, in some cases, how it will work.

In order to interpret the designer's or engineer's plans accurately, modelmakers must be able to read and understand drawings. They also must be able to prepare accurate drawings.

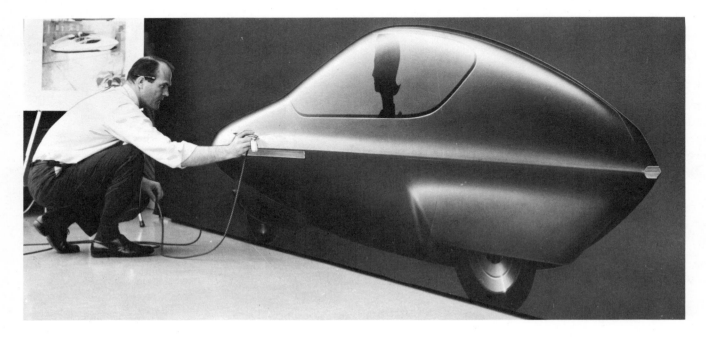

Fig. 26-5. After the ideas for a new car are carried through full-size line drawings, a full-size colored rendering is made to enable the designers to evaluate the shape as it would actually appear in paint and chrome. Here a technical illustrator uses an airbrush to put the finishing touches on a rendering of an experimental small car. (General Motors Corp.)

TECHNICAL ILLUSTRATORS

Technical illustration is a process of preparing art work for industry. The TECHNICAL ILLUSTRATOR prepares pictorial matter for engineering and educational purposes. See Fig. 26-5.

Technical illustrators must have technical and artistic abilities. The technical ability is necessary to understand the mechanical aspects of the job. The artistic ability will show the building or product in three dimensions. A technical illustrator must be able to make accurate sketches and finish the drawing according to industry standards.

TEST YOUR KNOWLEDGE - UNIT 26

(Write answers on a separate sheet of paper.)

1. Drafters make _____.
2. The drafter usually starts as a _____ and advances to _____ and _____ on the way to becoming lead or chief drafter.
3. A good drafter is _____, _____ and _____.
4. The industrial designer's main job is to _____.
5. _____ is a profession that offers a freedom not usually found in other professions.
6. _____in general, plan and design all kinds of structures and buildings. Most states require _____.
7. The engineer usually specializes in one of the many branches of the profession. List four kinds of engineering:
   a. _____.
   b. _____.
   c. _____.
   d. _____.
8. The modelmaker makes _____ to show the client _____.
9. The _____ must have a combination of technical and artistic abilities.

OUTSIDE ACTIVITIES

1. Invite a representative from the local Government Employment Service Office to discuss, with your class, employment opportunities in the drafting occupations.
2. Make a study of the HELP WANTED COLUMNS in your daily newspaper for a period of two weeks. Prepare a list of drafting and related jobs available, salaries offered and the minimum requirements for securing the jobs. Also, how often are additional benefits such as insurance, hospitalization, etc., mentioned in the ads?
3. Summarize the information on the drafting occupations given in the OCCUPATIONAL OUTLOOK HANDBOOK (a Government publication) and make it available to the class.

Unit 27

THE CHANGEOVER FROM CONVENTIONAL MEASUREMENT TO METRIC MEASUREMENT

The changeover from conventional measurement (inch, foot, yard, etc.) to metric measurement (millimetre, centimetre, metre, etc.) will not be easy. It will be very expensive and take many years.

Fig. 27-1. This four cylinder, 2300 cubic centimetre (2.3 litre) is the first metric powerplant produced in the United States. It was designed and manufactured entirely to metric standards by Ford Motor Company.

What problems can YOU foresee during the changeover? Since the change will be gradual, how will it be handled in the design and manufacture of new and replacement products? In the automotive industry, for example, the changeover has already started. See Fig. 27-1.

One problem for engineers and drafters will be at the INTERFACE (surfaces where parts come together) of a part designed to metric standards that must fit to an existing part that was designed to inch standards.

With this in mind, five methods have been devised for drafters to provide the information necessary to make the part.
1. Dual dimensioning.
2. Dimensioning with letters and tabular chart.
3. Metric dimensioning with readout chart.
4. Dimensioning with metric units only.
5. Undimensioned master drawings.

NOTE: In the following methods of presenting inch/millimetre dimensions, thread size used in the part remains to inch standards. There is no metric thread that is interchangeable with the thread size used.

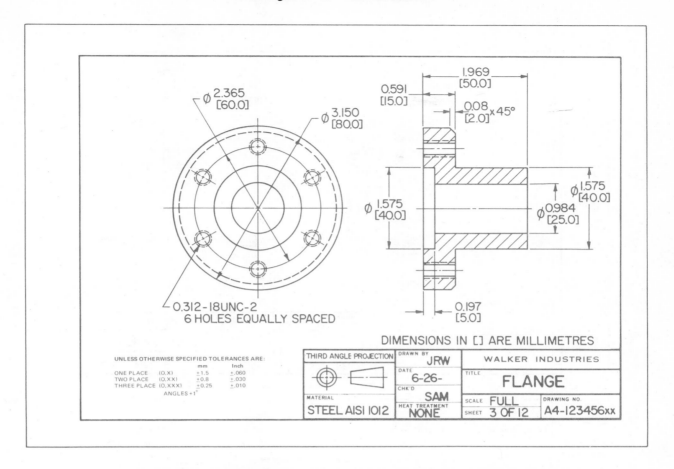

Fig. 27-2. DUAL DIMENSIONED drawing. Dual dimensioning has limited use. It was the first method devised to dimension engineering drawings with both inch and metric units. Notice important points on drawing (arrows): 1. NOTE indicating how metric dimensions are identified. 2. Thread size is NOT given in metric units because there is NO metric thread this size.

DUAL DIMENSIONING

The use of DUAL DIMENSIONING, Fig. 27-2, was the first method devised to dimension engineering drawings with both inch and metric units.

The dual dimensions were presented by the POSITION METHOD or the BRACKET METHOD, Fig. 27-3. The inch dimension was placed first on drawings for products to be made in the United States.

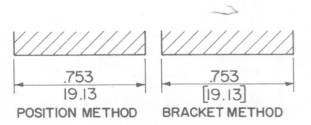

Fig. 27-3. Methods of indicating inches and millimetres on a dual dimensioned drawing. Avoid using BOTH methods on the same drawing.

The metric dimension was placed first on drawings of products to be produced where metrics was the basic form of measurement.

Dual dimensioning is the most complicated dimensioning system. IT IS NOT RECOMMENDED AS A PRACTICE FOR IMPLEMENTING THE TRANSITION FROM INCH TO METRIC UNITS.

Dual dimensioning is being replaced by other inch/mm dimensioning practices, some of which are detailed in the following paragraphs.

DIMENSIONING WITH LETTERS

Dimensioning with letters, Fig. 27-4, is a technique that is finding limited use. A letter (A, B, C, etc.) is used in place of either inch or millimetre dimensions. A TABULAR CHART is added to the drawing. The chart shows the metric and inch equivalents of each letter.

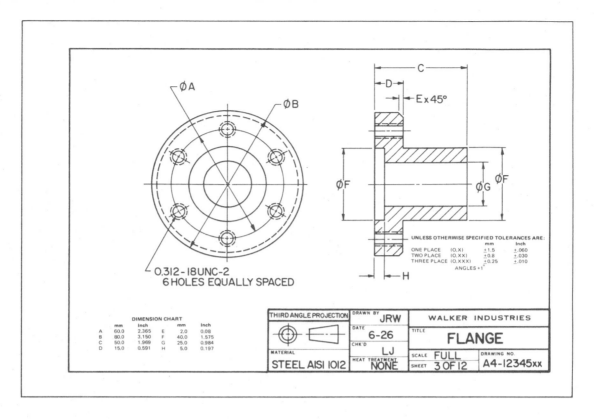

Fig. 27-4. Dimensioning with letters. Letters (A, B, C, etc.) are used on the drawing in place of either inch or millimetre dimensions. A TABULAR CHART at lower left of drawing shows the millimetre and inch equivalents for each letter.

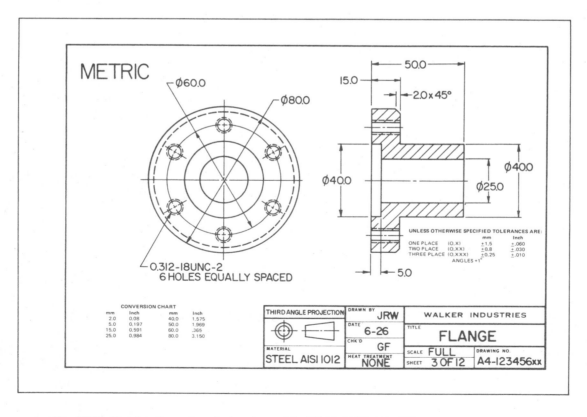

Fig. 27-5. Metric dimensioned drawing with READOUT CHART that shows the inch equivalents to the metric dimensions.

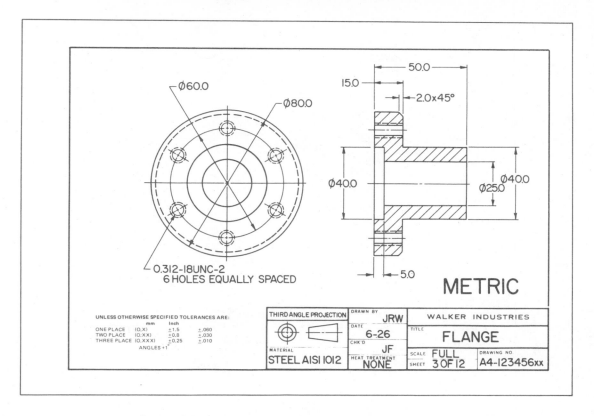

Fig. 27-6. Drawing dimensioned only in metric units. However, since there is NO metric equivalent for the thread indicated on the drawing, thread size is given in inch units.

METRIC DIMENSIONING WITH READOUT CHART

In doing metric dimensioning with READOUT CHART, the part is designed to metric standards. Only metric dimensions are placed on the drawing, Fig. 27-5. A readout chart is added to the drawing. It shows metric dimensions in the left column and the inch equivalents in the right column. This technique permits a comparison of values.

DIMENSIONING WITH METRIC UNITS ONLY

Dimensioning with metric units only is the quickest way to get drafting personnel and craftworkers to "think metric". Only metric dimensions are used on the drawing. See Fig. 27-6.

UNDIMENSIONED MASTER DRAWINGS

First, a MASTER DRAWING is made without dimensions, Fig. 27-7. Next, prints are made. The metric dimensions may be added to one print, inch dimensions to another print. Notes and details can be added in whatever language is needed to produce the part.

Whatever dimensioning technique is used, drafting personnel must have both metric and inch based scales and templates on hand, along with a thorough knowledge of the metric system.

Drafters, engineers or designers, for example, will find they cannot specify a 9/16 bolt by merely listing the metric equivalent of 9/16 (14.29 mm). THIS IS BECAUSE THERE ARE NO METRIC BOLTS THAT CORRESPOND TO THIS DIAMETER.

Various metric references can be found in Figs. 27-8 and 27-9.

The same problem will occur if a 1.0 inch diameter is specified as a 25.4 mm diameter. There is NO standard SI unit shaft of this diameter. The closest size would be 25.0 mm diameter. To get the 25.4 mm diameter shaft, an expensive machining operation would be necessary to turn a larger shaft to the 25.4 mm diameter.

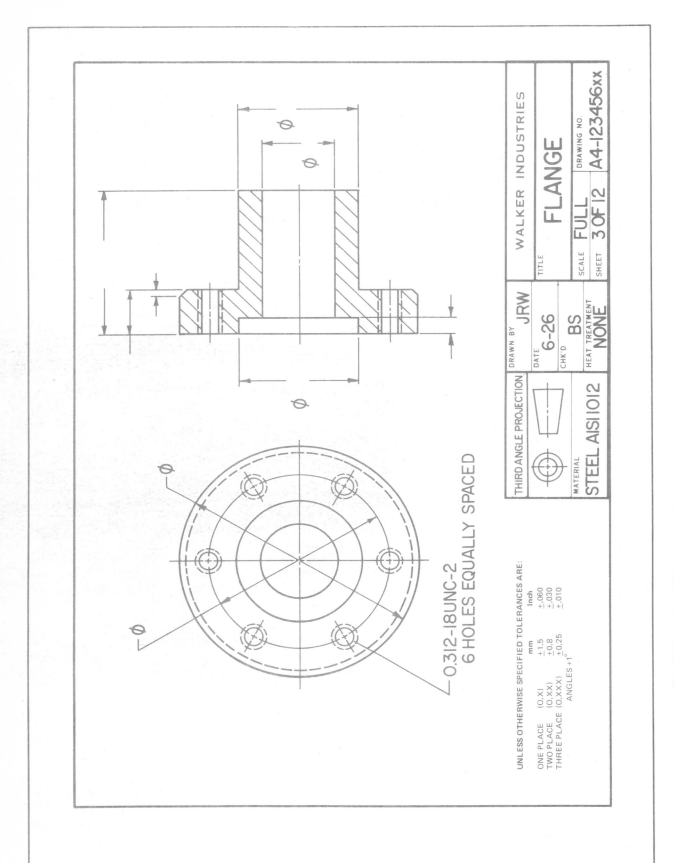

0.312-18UNC-2
6 HOLES EQUALLY SPACED

UNLESS OTHERWISE SPECIFIED TOLERANCES ARE:

	mm	Inch
ONE PLACE (0.X)	±1.5	±.060
TWO PLACE (0.XX)	±0.8	±.030
THREE PLACE (0.XXX)	±0.25	±.010
ANGLES ±1°		

THIRD ANGLE PROJECTION

WALKER INDUSTRIES

DRAWN BY JRW

DATE 6-26

CHK'D BS

TITLE FLANGE

SCALE FULL

SHEET 3 OF 12

DRAWING NO. A4-123456xx

HEAT TREATMENT NONE

MATERIAL STEEL AISI 1012

Fig. 27-7. A MASTER DRAWING. No dimensions are shown on the drawing (except for thread size). After the print is made of the drawing, dimensions, either inch or metric units, are added to the print.

SI UNITS & CONVERSIONS

PROPERTY	UNIT	SYMBOL	EXACT CONVERSION FROM	TO	MULTIPLY BY	APPROXIMATE EQUIVALENCY
length	metre	m	inch	mm	2.540×10	25mm = 1 in.
	centimetre	cm	inch	cm	2.540	300mm = 1 ft.
	millimetre	mm	foot	mm	3.048×10^{-4}	
mass	kilogram	kg	ounce	g	2.835×10	2.8g - 1 oz.
	gram	g	pound	kg	4.536×10^{-1}	kg = 2.2 lbs. = 35 oz.
	tonne (megagram)	t	ton (2000 lb)	kg	9.072×10^{2}	1t = 2200 lbs.
density	kilogram per cub. metre	kg/m^3	pounds per cu. ft.	kg/m^3	1.602×10	16kg/M^3 = 1 lb./ft^3
temperature	deg. Celsius	°C	deg. Fahr.	°C	$(°F-32) \times 5/9$	0°C = 32°F; 100°C = 212°F
area	square metre	m^2	sq. inch	mm^2	6.452×10^{2}	645mm^2 = 1 in.^{2}
	square millimetre	mm^2	sq. ft.	m^2	9.290×10^{-2}	1m^2 = 11 ft.^{2}
volume	cubic metre	m^3	cu. in.	mm^3	1.639×10^{4}	16400mm^3 = 1 in.^{3}
	cubic centimetre	cm^3	cu. ft.	m^3	2.832×10^{-2}	1m^3 = 35 ft.^{3}
	cubic millimetre	mm^3	cu. yd.	m^3	7.645×10^{-1}	1m^3 = 1.3 yd.^{3}
force	newton	N	ounce (Force)	N	2.780×10^{-1}	1N = 3.6 oz.
	kilonewton	kN	pound (Force)	kN	4.448×10^{-3}	4.4N = 1 lb.
	meganewton	MN	Kip	MN	4.448	1kN = 225 lb.
stress	megapascal	MPa	pound/in^2 (psi)	MPa	6.895×10^{-3}	1MPa = 145 psi
			Kip/in^2 (ksi)	MPa	6.895	7MPa = 1 ksi
torque	newton-metres	N.m	in-ounce	N.m	7.062×10^{3}	1N.m = 140 in.oz.
			in.pound	N.m	1.130×10^{-1}	1N.m = 9 in.lb.
			ft pound	N.m	1.356	1N.m = .75 ft.lb.
						1.4N.m = 1 ft.lb.

Fig. 27-8. Metric units and conversions.

TWIST DRILL DATA

METRIC DRILL SIZES (mm)[1]		Decimal Equivalent in Inches (Ref)
Preferred	Available	
	.40	.0157
	.42	.0165
	.45	.0177
	.48	.0189
.50		.0197
	.52	.0205
.55		.0217
	.58	.0228
.60		.0236
	.62	.0244
.65		.0256
	.68	.0268
.70		.0276
	.72	.0283
.75		.0295
	.78	.0307
.80		.0315
	.82	.0323
.85		.0335
	.88	.0346
.90		.0354
	.92	.0362
.95		.0374
	.98	.0386
1.00		.0394
	1.03	.0406
1.05		.0413
	1.08	.0425
1.10		.0433
	1.15	.0453
1.20		.0472
1.25		.0492
1.30		.0512
	1.35	.0531
1.40		.0551
	1.45	.0571
1.50		.0591
	1.55	.0610
1.60		.0630
	1.65	.0650

1 Metric drill sizes listed in the "Preferred" column are based on the R'40 series of preferred numbers shown in the ISO Standard R497. Those listed in the "Available" column are based on the R80 series from the same document.

TWIST DRILL DATA (CONTINUED)

METRIC DRILL SIZES (mm)[1]		Decimal Equivalent in Inches (Ref)
Preferred	Available	
6.00		.2362
	6.20	.2441
6.30		.2480
	6.50	.2559
6.70		.2638
	6.80[2]	.2677
	6.90	.2717
7.10		.2795
	7.30	.2874
7.50		.2953
	7.80	.3071
8.00		.3150
	8.20	.3228
8.50[2]		.3346
	8.80	.3465
9.00		.3543
	9.20	.3622
9.50		.3740
	9.80	.3858
10.00		.3937
	10.30	.4055
10.50		.4134
	10.80	.4252
11.00		.4331
	11.50	.4528
12.00		.4724
12.50		.4921
13.00		.5118
	13.50	.5315
14.00		.5512
	14.50	.5709
15.00		.5906
	15.50	.6102
16.00		.6299
	16.50	.6496
17.00		.6693
	17.50	.6890
18.00		.7087
	18.50	.7283
19.00		.7480

METRIC DRILL SIZES (mm)[1]		Decimal Equivalent in Inches (Ref)
Preferred	Available	
	19.50	.7677
20.00		.7874
	20.50	.8071
21.00		.8268
	21.50	.8465
22.00		.8661
	23.00	.9055
24.00		.9449
25.00		.9843
26.00		1.0236
	27.00	1.0630
28.00		1.1024
	29.00	1.1417
30.00		1.1811
	31.00	1.2205
32.00		1.2598
	33.00	1.2992
34.00		1.3386
	35.00	1.3780
36.00		1.4173
	37.00	1.4567
38.00		1.4961
	39.00	1.5354
40.00		1.5748
	41.00	1.6142
42.00		1.6535
	43.50	1.7126
45.00		1.7717
	46.50	1.8307
48.00		1.8898
50.00		1.9685
	51.50	2.0276
53.00		2.0866
	54.00	2.1260
56.00		2.2047
	58.00	2.2835
60.00		2.3622

1 Metric drill sizes listed in the "Preferred" column are based on the R'40 series of preferred numbers shown in the ISO Standard R497. Those listed in the "Available" column are based on the R80 series from the same document.
2 Recommended only for use as a tap drill size.

Fig. 27-9. Metric twist drill data. (General Motors)

TEST YOUR KNOWLEDGE - UNIT 27

(Write answers on a separate sheet of paper.)

1. What problems do YOU foresee in the changeover from the inch based system of measurement to metrics?
2. List the five (5) dimensioning methods recommended for use during the changeover.
   a. _____.
   b. _____.
   c. _____.
   d. _____.
   e. _____.
3. Give a brief description of each of the above methods.
4. What is a MASTER DRAWING?

OUTSIDE ACTIVITIES

1. Secure examples of the five dimensioning methods recommended for use during the changeover from inch based measurement to metrics.

GLOSSARY

ACCURATE: Made within the tolerances allowed.

ACOUSTIC TILE: Tile made of sound absorbing materials.

ACUTE ANGLE: An angle less than 90 deg.

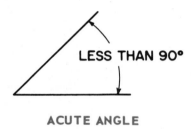

ACUTE ANGLE

AIR BRUSH: A device used to spray paint by means of compressed air.

AIR CONDITIONING: The control of the temperature, humidity, motion, dust and distribution of the air within a structure.

ALLOY: A mixture of two or more metals fused or melted together to form a new metal.

ALLOWANCE: The limits permitted for satisfactory performance of machined parts.

ANGLE: The figure formed by two lines coming together to a point.

ANNEALING: The process of heating metal to a given temperature and cooling it slowly to remove stresses and induce softness. The exact temperature and period the temperature is held depend upon the composition of the metal being annealed.

ARC: Portion of a circle.

ASPHALT ROOFING: A roofing material made by saturating an asbestos or felt with asphalt.

ASSEMBLY: A unit fitted together from manufactured parts.

ATTIC: The part of a building just below the roof.

AXES: The plural of axis.

AXIS: The center line of a view or of a geometric figure.

BASEMENT: The base story of a structure. It usually is underground.

BEAM: A term used to describe joists, rafters and girders.

BEAM COMPASS: Compass used to draw large circles and arcs.

BEVEL: The angle that is formed by a line or a surface that is not at right angles to another line or surface.

BLOWHOLE: A hole produced in a casting when gases are entrapped during the pouring operation.

BLUEPRINT: A reproduction of a drawing that has a bright background with white lines.

BRAZING: Joining metals by the fusion of nonferrous alloys that have melting temperatures above 800 deg. F., but lower than the metals being joined.

BUSHING: A bearing for a revolving shaft. Also, a hardened steel tube used on jigs to guide drills and reamers.

CASTING: An object made by pouring molten metal into a mold.

CAULK: To seal cracks and seams with a waterproofing material.

CASEHARDEN: A process of surface hardening iron base alloys so that the surface layer or case is made substantially harder than the interior or core of the metal.

CASEHARDENING

CENTER LINE SYMBOL.

CHAMFER: See BEVEL.
CIRCUMFERENCE: The perimeter of a circle.
CIRCUMSCRIBE: To draw a line around.
CLEARANCE: The distance by which one part clears another part.
CLOCKWISE: From left to right in a circular motion. The direction clock hands move.

CLOCKWISE

COMPUTER: An electronic calculator performing a sequence of computations.
CONCAVE SURFACE: A curved depression in the surface of an object.

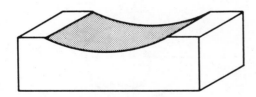

CONCAVE SURFACE

CONCENTRIC: Having a common center.

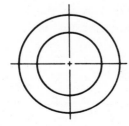

CONCENTRIC CIRCLES

CONCRETE: A mixture of portland cement, sand, gravel and water.
CONICAL: Shaped like a cone.

CONTOUR: The outline of an object.
CONVENTIONAL: Customary or traditional; not original.
CONVEX SURFACE: A rounded surface on an object.

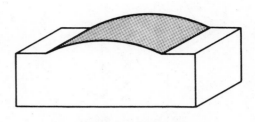

CONVEX SURFACE

CORE: A body of sand or other material that is formed to a desired shape and placed in a mold to produce a cavity or opening in a casting.
COUNTERBORE: To enlarge a hole to a given depth.

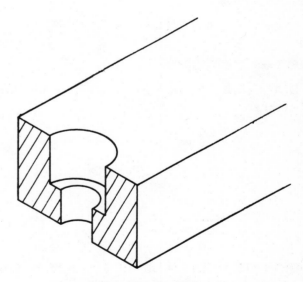

COUNTERBORED HOLE

COUNTERCLOCKWISE: From right to left in a circular motion.

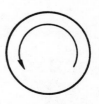

COUNTERCLOCKWISE

COUNTERSINK: To chamfer a hole to receive a flat or fillister head fastener.

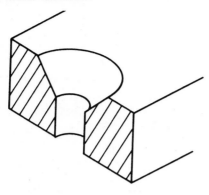

COUNTERSUNK HOLE

CURVED LINE: A line of which no part is straight.

CYLINDER: A geometric figure with a uniform circular cross section through its entire length.

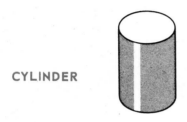

CYLINDER

DIAGONAL: A line running across from corner to corner.

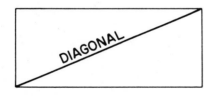

DIAMETER: The length of a straight line running through the center of a circle.

DIE: The tool used to cut external threads. Also, a tool used to shape materials.

DIE CASTING: A method of casting metal under pressure by injecting it into metal dies of a die casting machine.

DRAFT: The clearance on a pattern or mold that allows easy withdrawal of pattern from the mold.

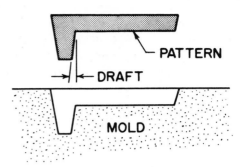

DRILLING: Cutting round holes by use of a cutting tool called a drill.

DRILL ROD: Accurately ground and polished tool steel rods.

DRIVE FIT: Using force or pressure to fit two pieces together. Also, one of several classes of fits.

DUCT: A sheet metal pipe or passageway used for conveying air.

ECCENTRIC: Not on a common center.

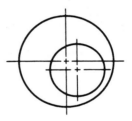

ECCENTRIC CIRCLES

ELIMINATE: To do away with.

ELLIPSE: A closed curve in the form of a symmetrical oval.

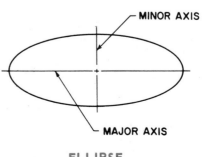

ELLIPSE

ENGINEERING DRAWING: The graphic language (technical drawing or drafting) of the engineer.

EQUAL: The same.

EQUILATERAL: A figure having equal length sides.

EXPANSION FIT: The reverse of shrink fit. The piece to be fitted is placed in liquid nitrogen or dry ice until it shrinks enough to fit into the mating piece. Interference develops between the fitted pieces as the cooled piece expands to normal size.

FERROUS METAL: A metal that contains iron as its major ingredient.

FILLET: The curved surface connecting two surfaces that form an angle.

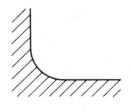

FILLET

FIXTURE: A device for holding metal while it is being machined.

FORGE: To form material, using heat and pressure.

FLASH: A thin fin of metal formed at the parting line of a forging or casting where a small portion of metal is forced out between the edges of the die.

FLASK: A frame of wood or metal consisting of a cope (the top portion) and a drag (the bottom portion) used to hold sand that forms the mold used in the foundry.

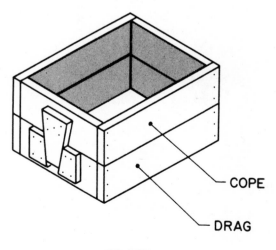

COPE

DRAG

FLASK

FLOOR PLAN: The drawing that shows the exact shape, dimensions and arrangements of the rooms of a building.

FORCE FIT: The interference between two mating parts sufficient to require force to press the pieces together. The joined pieces are considered permanently assembled.

FREE FIT: Used when tolerances are liberal. Clearance is sufficient to permit a shaft to turn freely without binding or overheating when properly lubricated.

FRUSTUM: The figure formed by cutting off a portion of a cone or pyramid parallel to its base.

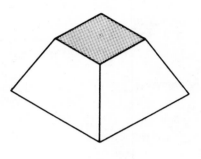

FRUSTUM

GATE: The opening that guides the molten metal into the cavity that forms the mold. See SPRUE.

GEARS: Toothed wheels that transmit rotary motion from one shaft without slippage.

HARDENING: The heating and quenching of certain iron-base alloys for the purpose of producing a hardness superior to that of the untreated metal.

HEAT TREATMENT: The careful application of a combination of heating and cooling cycles to a metal or alloy to bring about certain desirable conditions such as hardness and toughness.

HEXAGON: A six-sided figure with each side forming a 60 deg. angle.

INCLINED: Making an angle with another line or plane.

INSCRIBE: To draw one figure within another figure.

INSPECTION: The measuring and checking of finished parts to determine whether they have been made to specifications.

INTERCHANGEABLE: Refers to a part that has been made to specific dimensions and tolerances and is capable of being fitted in a mechanism in place of a similarly made part.

INVESTMENT CASTING: A process that involves making a wax, plastic or a frozen mercury pattern and surrounding it with a wet refractory material. After the investment material has dried and set, the pattern is melted or burned out and molten metal is poured into the cavity.

JAMB: The vertical parts of a door frame.

JIG: A device that holds the work in position and guides the cutting tool.

KEY: A small piece of metal imbedded partially in the shaft and partially in the hub to prevent rotation of a gear or pulley on the shaft.

KEYWAY: The slot or recess in the shaft that holds the key.

LAY OUT: To locate and scribe points for machining and forming operations.

LINE CONVENTIONS: Symbols that furnish a means of representing or describing some part of an object. It is expressed by a combination of line weight and appearance.

MACHINE TOOL: The name given to that class of machines which, taken as a group, can reproduce themselves.

MAJOR DIAMETER: The largest diameter of a thread measured perpendicular to the axis.

MESH: To engage gears to a working contact.

MICROINCH: One-millionth of an inch.

MILL: To remove metal with a rotating cutter on a milling machines.

MINOR DIAMETER: The smallest diameter on a screw thread measured across the root of the thread and perpendicular to the axis. Also known as the "root diameter."

MOLD: The material that forms the cavity into which molten metal is poured.

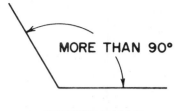

OBTUSE ANGLE

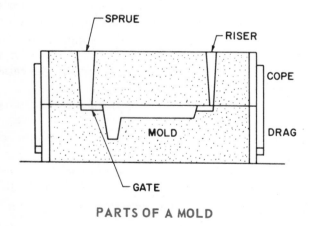

PARTS OF A MOLD

NATIONAL COARSE (NC): The coarse series of the American National thread series.

NATIONAL FINE (NF): The fine series of the American National thread series.

NC: Abbreviation for the National Coarse series of threads.

NF: Abbreviation for the National Fine series of threads.

OBTUSE ANGLE: An angle more than 90 deg.

OCTAGON: An eight-sided geometric figure with each side forming a 45 deg. angle.

OD: Abbreviation for outside diameter.

PENTAGON: A five-sided geometric figure with each side forming a 72 deg. angle.

PERIMETER: The boundry of a geometric figure.

PERMANENT MOLD: Mold ordinarily made of metal that is used for the repeated production of similar castings.

PERPENDICULAR: A line at right angles to a given line.

PHOTO DRAWING: A drawing prepared using a photograph on which dimensions, notes and specifications have been added.

ROUGH LAYOUT: A rough pencil plan that arranges lines and symbols so that they have a pleasing relation to one another.

PITCH: The distance from a point on one thread to a corresponding point on the next thread.

PLATE: Another name given to a drawing.

PROFILE: The outline of an object.

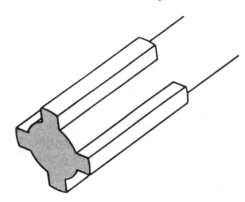

PROFILE

PROJECT: To extend from.

RACK: A flat strip with teeth designed to mesh with teeth on a gear. Used to change rotary motion to a reciprocating motion.

RADIUS: The length of a straight line running from the center of a circle to the perimeter of the circle.

RADII: The plural of radius.

REAM: To finish a drilled hole to exact size with a reamer.

RECTANGLE: A geometric figure with opposite sides equal in length and each corner forming a 90 deg. angle.

RIGHT ANGLE: A 90 deg. angle. The angle that is formed by a line that is perpendicular to another line.

RISER: An opening in the mold that permits the gases to escape. The gases are formed when molten metal is poured into the mold.

ROTATE: To turn or revolve around a point.

SEAM: The line formed where two edges are joined together.

SEGMENT: Any part of a divided line.

SKETCH: To draw without the aid of drafting instruments.

SLIDE RULE: A portable calculating device.

SPLINE: A series of grooves, cut lengthwise, around a shaft or hole.

SPOTFACE: To machine a circular spot on the surface of a casting to furnish a bearing surface for the head of a bolt or a nut.

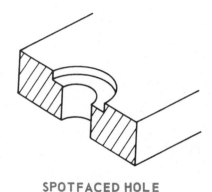

SPOTFACED HOLE

SPRUE: The opening in a mold that leads to the gate which, in turn, leads to the cavity into which molten metal is poured.

SQUARE: To machine or cut at right angles. Also, a geometric figure with four equal length sides and four right (90 deg.) angles.

SYMBOL: A figure or character used in place of a word or group of words.

TAP: The tool used to cut internal threads.

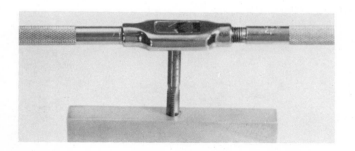

TAPER: A piece that increases or decreases in size at a uniform rate to assume a wedge or conical shape.

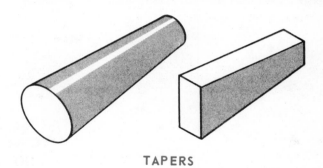

TAPERS

TEMPLATE: A pattern or guide.

THREAD: The act of cutting a screw thread.

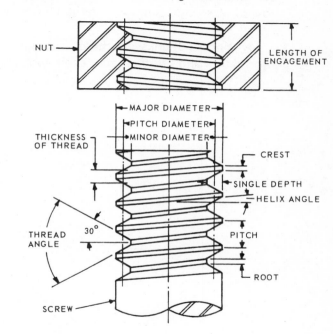

THREADS

TRAIN: A series of meshed gears.

TRIANGLE: A three-sided geometric figure.

TRUNCATE: To cut off a geometric solid at an angle to its base.

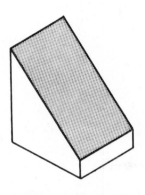

TRUNCATED PRISM

UNIFIED THREADS: A series of screw threads that have been adopted by the United States, Canada and Great Britain to attain interchangeability of certain screw threads.

VERTICAL: At right angles to a horizontal line or plane.

WORKING DRAWING: A drawing that gives information needed to make and assemble a product.

ACKNOWLEDGMENTS

While it would be a most pleasant task, it would be impossible for one person to develop the material included in this text by visiting the various industries represented and observing, studying and taking the photos first hand.

My sincere thanks to those who helped in the gathering of the necessary material, information and photographs. Their cooperation was most appreciated.

John R. Walker
Bel Air, Maryland

TABLES

METRIC SIZE DRAFTING SHEETS

ISO STANDARD

SIZE	MILLIMETRES	INCHES
AO	841 x 1189	33.11 x 46.81
A1	594 x 841	23.39 x 33.11
A2	420 x 594	16.54 x 23.39
A3	297 x 420	11.69 x 16.54
A4	210 x 297	8.27 x 11.69

AMERICAN STANDARD

SIZE	INCHES	MILLIMETRES
E	34 x 44	863.6 x 1117.6
D	22 x 34	558.8 x 863.6
C	17 x 22	431.8 x 558.8
B	11 x 17	279.4 x 431.8
A	8 1/2 x 11	215.9 x 279.4

Left. Metric size drafting paper sheets are exactly proportional in size. Right. Charts show millimetre sizes and customary inch equivalent sizes of ISO papers.

ABBREVIATIONS

(SOME FORMER ABBREVIATIONS ARE NOW SYMBOLS)

Across flats	ACR FLT	Inside diameter	ID
Centers	CTR	Left hand	LH
Center line	CL	Material	MATL
Centimetre	cm	Metre	m
Chamfer	CHAM	Millimetre	mm
Counterbore	CBORE	Number	NO
Countersink	CSK	Outside diameter	OD
Countersunk head	CSK H	Pitch diameter	PD
Diameter (before dimension)	Ø	Radius	R
Diameter (in a note)	DIA	Right hand	RH
Drawing	DWG	Round	RD
Figure	FIG	Square (before dimension)	□
Hexagon	HEX	Square (in a note)	SQ
Hexagonal head	HEX HD	Thread	THD

Exploring Metric Drafting

INCH	DECIMAL INCH	MILLIMETRE	INCH	DECIMAL INCH	MILLIMETRE
1/64	0.0156	0.3967	33/64	0.5162	13.0968
1/32	0.0312	0.7937	17/32	0.5312	13.4937
3/64	0.0468	1.1906	35/64	0.5468	13.8906
1/16	0.0625	1.5875	9/16	0.5625	14.2875
5/64	0.0781	1.9843	37/64	0.5781	14.6843
3/32	0.0937	2.3812	19/32	0.5937	15.0812
7/64	0.1093	2.7781	39/64	0.6093	15.4781
1/8	0.125	3.175	5/8	0.625	15.875
9/64	0.1406	3.5718	41/64	0.6406	16.2718
5/32	0.1562	3.9687	21/32	0.6562	16.6687
11/64	0.1718	4.3656	43/64	0.6718	17.0656
3/16	0.1875	4.7625	11/16	0.6875	17.4625
13/64	0.2031	5.1593	45/64	0.7031	17.8593
7/32	0.2187	5.5562	23/32	0.7187	18.2562
15/64	0.2343	5.9531	47/64	0.7343	18.6531
1/4	0.25	6.5	3/4	0.75	19.05
17/64	0.2656	6.7468	49/64	0.7656	19.4468
9/32	0.2812	7.1437	25/32	0.7812	19.8437
19/64	0.2968	7.5406	51/64	0.7968	20.2406
5/16	0.3125	7.9375	13/16	0.8125	20.6375
21/64	0.3281	8.3343	53/64	0.8281	21.0343
11/32	0.3437	8.7312	27/32	0.8437	21.4312
23/64	0.3593	9.1281	55/64	0.8593	21.8281
3/8	0.375	9.525	7/8	0.875	22.225
25/64	0.3906	9.9218	57/64	0.8906	22.6218
13/32	0.4062	10.3187	29/32	0.9062	23.0187
27/64	0.4218	10.7156	59/64	0.9218	23.4156
7/16	0.4375	11.1125	15/16	0.9375	23.8125
29/64	0.4531	11.5093	61/64	0.9531	24.2093
15/32	0.4687	11.9062	31/32	0.9687	24.6062
31/64	0.4843	12.3031	63/64	0.9843	25.0031
1/2	0.50	12.7	1	1.0000	25.4

Tables

SIZES 1 mm to 10 mm			SIZES 10 mm to 100 mm			SIZES 100 mm to 1000 mm		
CHOICE			CHOICE			CHOICE		
1st	2nd	3rd	1st	2nd	3rd	1st	2nd	3rd
1			10			100		
								105
	1.1			11		110		
								115
1.2			12			120		
								125
		1.3			13		130	
								135
	1.4			14		140		
								145
		1.5			15		150	
								155
1.6			16			160		
								165
		1.7			17		170	
								175
	1.8			18		180		
								185
		1.9			19		190	
								195
2			20			200		
		2.1			21		210	
	2.2			22		220		
					23		230	
		2.4			24		240	
2.5			25			250		
		2.6			26		260	
								270
	2.8			28		280		
								290
3			30			300		
		3.2		32			320	
					34			340
	3.5			35		350		
					36			360
		3.8		38			380	
4			40			400		
		4.2		42			420	
					44			440
	4.5			45		450		
					46			460
		4.8		48			480	
5			50			500		
		5.2		52			520	
					54			540
	5.5			55		550		
					56			560
		5.8		58			580	
6			60			600		
					62		620	
								640
	6.5			65		650		
								660
					68		680	
	7			70		700		
					72		720	
								740
	7.5			75		750		
								760
					78		780	
8			80			800		
					82			820
	8.5			85		850		
					88			880
	9			90		900		
					92			920
		9.5		95			950	
					98			980
10			100			1000		

INCH SERIES			METRIC			
Size	Dia. in.	TPI	Size	Dia. in.	Pitch (mm)	TPI (Approx.)
			M1.4	.055	0.3	85
No. 0	.060	80				
			M1.6	.063	0.35(a)	74
No. 1	.073	64				
		72	M2	.079	0.4(a)	64
No. 2	0.86	56				
		64	M2.5	.098	0.45(a)	56
No. 3	.099	48				
		56				
No. 4	.112	40				
		48	M3	.118	0.5(a)	51
No. 5	.125	40				
		44				
No. 6	.138	32	M3.5	.138	0.6(a)	42
		40				
No. 8	.164	32	M4	.158	0.7(a)	36
		36				
No. 10	.190	24				
		32				
			M5	.197	0.8(a)	32
			M6	.236	1.0	25
1/4	.250	20				
		28	M6.3	.248	1.0(a)	25
5/16	.312	18				
		24	M7	.276	1.0	25
			M8	.315	1.25(a)	20
					1.0	25
3/8	.375	16				
		24				
			M10	.394	1.5(a)	18
					1.25	20
7/16	.437	14				
		20				
			M12	.472	1.75(a)	14.5
					1.25	20
1/2	.500	13				
		20				
9/16	.562	12	M14	.551	2.0(a)	12.5
		18			1.5	17
5/8	.625	11				
		18				
			M16	.630	2.0(a)	12.5
					1.5	17
			M18	.709	2.5	10
					1.5	17
3/4	.750	10				
		16	M20	.787	2.5(a)	10
					1.5	17
			M22	.866	2.5	10
					1.5	17
7/8	.875	9				
		14	M24	.945	3.0(a)	8.5
					2.0	12.5
1	1.000	8				
		14	M27	1.063	3.0	8.5
					2.0	12.5

(a) Preferred series for use in the United States

A comparison of inch series and metric threads shows that even though many of the threads are similar in size and pitch they are not interchangeable. Extreme care must be observed to keep the two series separate.

Exploring Metric Drafting

NOMINAL SIZE	INTERNAL THREAD MINOR DIAMETER		TAP DRILL DIAMETER
	Max	Min	
M1.6 x 0.35	1.321	1.221	1.25
M2 x 0.4	1.679	1.567	1.6
M2.5 x 0.45	2.138	2.013	2.05
M3 x 0.5	2.599	2.459	2.5
M3.5 x 0.6	3.010	2.850	2.9
M4 x 0.7	3.422	3.242	3.3
M5 x 0.8	4.334	4.134	4.2
M6.3 x 1	5.553	5.217	5.3
M8 x 1.25	6.912	6.647	6.8
M10 x 1.5	8.676	8.376	8.5
M12 x 1.75	10.441	10.106	10.2
M14 x 2	12.210	11.835	12.0
M16 x 2	14.210	13.835	14.0
M20 x 2.5	17.744	17.294	17.5
M24 x 3	21.252	20.752	21.0
M30 x 3.5	26.771	26.211	26.5
M36 x 4	32.270	31.670	32.0
M42 x 4.5	37.799	37.129	37.5
M48 x 5	43.297	42.587	43.0
M56 x 5.5	50.796	50.046	50.5
M64 x 6	58.305	57.505	58.0
M72 x 6	66.305	65.505	66.0
M80 x 6	74.305	73.505	74.0
M90 x 6	84.305	83.505	84.0
M100 x 6	94.305	93.505	94.0

The ISO series of 25 selected thread sizes that will be used by North American manufacturers.

CONVERSION TABLES

TO CONVERT	MULTIPLY BY	TO CONVERT	MULTIPLY BY
Length			
km to mi.	0.62	mi. to km	1.61
m to mi.	0.00062	mi. to m	1609.35
m to yd.	1.0936	yd. to m	0.9144
cm to in.	0.3937	in. to cm	2.54
mm to in.	0.03937	in. to mm	25.4
Volume			
cm^3 to cu. in.	0.061	cu. in. to cm^3 or ml	16.387
L to cu. in.	61.024	cu. in. to L	0.0164
L to gal.	0.264	gal. to L	3.785
Weight			
kg to lb.	2.2	lb. to kg	0.4536
g to oz.	0.0353	oz. to g	28.35

ABBREVIATIONS

mm	-	millimetre	km	-	kilometre
cm	-	centimetre	in.	-	inch
m	-	metre	ft.	-	foot
cm^3	-	cubic centimetre	yd.	-	yard
ml	-	millimetre	mi.	-	mile
L	-	litre	cu. in.	-	cubic inch
g	-	gram	oz.	-	ounce
kg	-	kilogram	gal.	-	gallon

METRIC SYSTEM

The basic unit of the metric system is the metre (m). The metre is exactly 39.37 in. long. This is 3.37 in. longer than the English yard. Units that are multiples or fractional parts of the metre are designated as such by prefixes to the work "metre". For example:

1 millimetre (mm)	= 0.001 metre or 1/1000 metre
1 centimetre (cm)	= 0.01 metre or 1/100 metre
1 decimetre (dm)	= 0.1 metre or 1/10 metre
	1 metre (m)
1 decametre (dkm)	= 10 metres
1 hectometre (hm)	= 100 metres
1 kilometre (km)	= 1000 metres

These prefixes may be applied to any unit of length, weight, volume, etc. The metre is adopted as the basic unit of length, the gram for mass, and the litre for volume.

In the metric system, area is measured in square kilometres (km^2), square centimetres (cm^2), etc. Volume is commonly measured in cubic centimetres, etc. One litre (L) is equal to 1,000 cubic centimetres.

The metric measurements in most common use are shown in the following tables:

MEASURES OF LENGTH

10 millimetres	= 1 centimetre
10 centimetres	= 1 decimetre
10 decimetres	= 1 metre
1000 metres	= 1 kilometre

MEASURES OF WEIGHT

100 milligrams	= 1 gram
1000 grams	= 1 kilogram
1000 kilograms	= 1 metric ton

MEASURES OF VOLUME

1000 cubic centimetres	= 1 litre
100 litres	= 1 hectolitre

FORMULAS

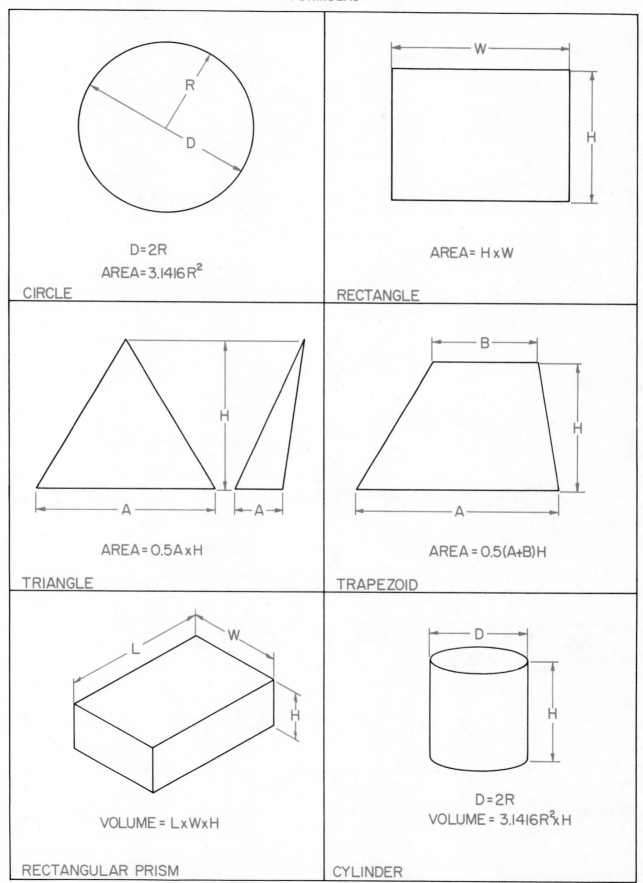

D=2R
AREA= 3.1416 R^2

CIRCLE

AREA= H x W

RECTANGLE

AREA= 0.5A x H

TRIANGLE

AREA = 0.5(A+B)H

TRAPEZOID

VOLUME = L x W x H

RECTANGULAR PRISM

D=2R
VOLUME = 3.1416 R^2 x H

CYLINDER

FORMULAS

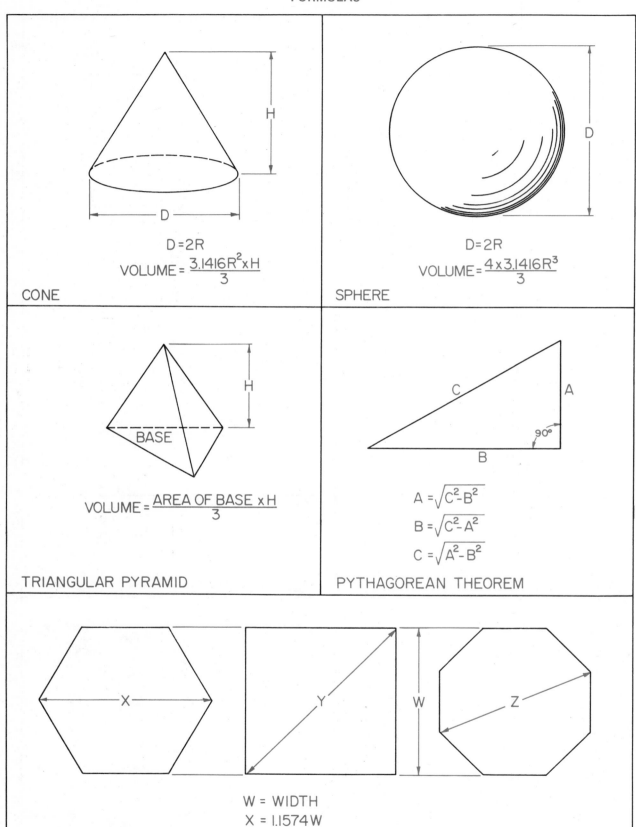

CONE

$D = 2R$

$$\text{VOLUME} = \frac{3.1416R^2 \times H}{3}$$

SPHERE

$D = 2R$

$$\text{VOLUME} = \frac{4 \times 3.1416R^3}{3}$$

TRIANGULAR PYRAMID

$$\text{VOLUME} = \frac{\text{AREA OF BASE} \times H}{3}$$

PYTHAGOREAN THEOREM

$$A = \sqrt{C^2 - B^2}$$

$$B = \sqrt{C^2 - A^2}$$

$$C = \sqrt{A^2 - B^2}$$

W = WIDTH
X = 1.1574 W
Y = 1.4142 W
Z = 1.0824 W

CALCULATING THE METRE

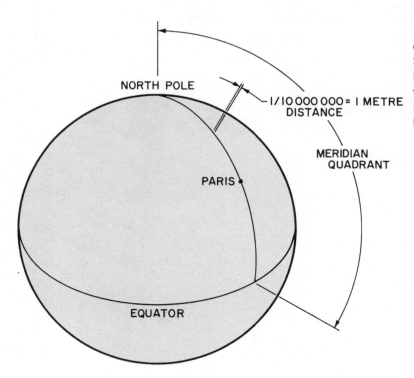

NORTH POLE

1/10 000 000 = 1 METRE DISTANCE

MERIDIAN QUADRANT

PARIS

EQUATOR

Fig. T-1. In 1793 the French government adopted a system of standards they called the METRIC SYSTEM. It was based on what they called a METRE. The metre was calculated to be one ten-millionth part of the distance from the North Pole to the Equator when measured on a straight line that ran along the surface of the earth through Paris.

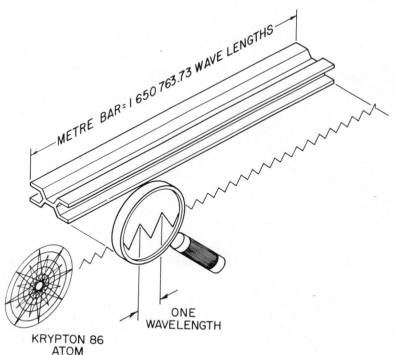

METRE BAR = 1 650 763.73 WAVE LENGTHS

ONE WAVELENGTH

KRYPTON 86 ATOM

Fig. T-2. The metre based on the measurement of the earth was not accurate for modern use. The modern metre is defined as 1 650 736.73 wavelengths in vacuum of the orange-red line of the spectrum of krypton-86. An INTERFEROMETER is used to measure the length by means of light waves. The reason the metre is defined in terms of the wavelength of light is so it can be reproduced in any modern scientific laboratory. By using a specific light wave, an accuracy of one part in a hundred million can be maintained.

PREFIXES, EXPONENTS AND SYMBOLS

DECIMAL FORM	EXPONENT OR POWER	PREFIX	PRONUNCIATION	SYMBOL	MEANING
1 000 000 000 000 000 000	$= 10^{18}$	exa	ex'a	E	quintillion
1 000 000 000 000 000	$= 10^{15}$	peta	pet'a	P	quadrillion
1 000 000 000 000	$= 10^{12}$	tera	tĕr'ả	T	trillion
1 000 000 000	$= 10^{9}$	giga	jǐ'gả	G	billion
1 000 000	$= 10^{6}$	mega	mĕg'ả	M	million
1 000	$= 10^{3}$	kilo	kĭl'ō	k	thousand
100	$= 10^{2}$	hecto	hĕk'to	h	hundred
10	$= 10^{1}$	deka	dĕk'a	da	ten
1					base unit
0.1	$= 10^{-1}$	deci	dĕs'ĭ	d	tenth
0.01	$= 10^{-2}$	centi	sĕn'tĭ	c	hundredth
0.001	$= 10^{-3}$	milli	mǐl'ĭ	m	thousandths
0.000 001	$= 10^{-6}$	micro	mi'krō	μ	millionth
0.000 000 001	$= 10^{-9}$	nano	nǎn'ō	n	billionth
0.000 000 000 001	$= 10^{-12}$	pico	pēc'ō	p	trillionth
0.000 000 000 000 001	$= 10^{-15}$	femto	fĕm'tō	f	quadrillionth
0.000 000 000 000 000 001	$= 10^{-18}$	atto	ǎt'tō	a	quintillionth

Most commonly used

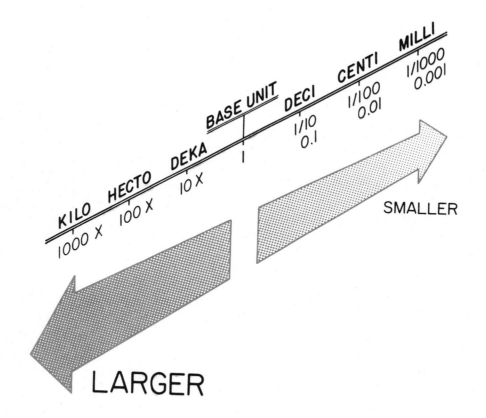

Tables

STANDARD ABBREVIATIONS
FOR USE ON DRAWINGS

A

Abrasive	ABRSV
Accessory	ACCESS
Accumulator	ACCUMR
Acetylene	ACET
Actual	ACT
Actuator	ACTR
Addendum	ADD
Adhesive	ADH
Adjust	ADJ
Advance	ADV
Aeronautic	AERO
Alclad	CLAD
Alignment	ALIGN
Allowance	ALLOW
Alloy	ALY
Alteration	ALT
Alternate	ALT
Alternating Current	AC
Aluminum	AL
American National Standards Institute	ANSI
American Wire Gage	AWG
Ammeter	AMM
Amplifier	AMPL
Anneal	ANL
Anodize	ANOD
Antenna	ANT
Approved	APPD
Approximate	APPROX
Arrangement	ARR
As Required	AR
Assemble	ASSEM
Assembly	ASSY
Automatic	AUTO
Auxiliary	AUX
Average	AVG

B

Babbit	BAB
Base Line	BL
Battery	BAT
Bearing	BRG
Bend Radius	BR
Bevel	BEV
Bill of Material	B/M
Blueprint	BP or B/P
Bolt Circle	BC
Bracket	BRKT
Brass	BRS
Brazing	BRZG

Brinnell Hardness Number	BHN
Bronze	BRZ
Brown & Sharpe (Gage)	B&S
Burnish	BNH
Bushing	BUSH

C

Cabinet	CAB
Calculated	CACL
Cancelled	CANC
Capacitor	CAP
Capacity	CAP
Carburize	CARB
Case Harden	CH
Casting	CSTG
Cast Iron	CI
Cathode-Ray Tube	CRT
Center	CTR
Center to Center	C to C
Centigrade	C
Centimeter	CM
Centrifugal	CENT
Chamfer	CHAM
Circuit	CKT
Circular	CIR
Circumference	CIRC
Clearance	CL
Clockwise	CW
Closure	CLOS
Coated	CTD
Cold-Drawn Steel	CDS
Cold-Rolled Steel	CRS
Color Code	CC
Commercial	COMM
Concentric	CONC
Condition	COND
Conductor	CNDCT
Contour	CTR
Control	CONT
Copper	COP
Counterbore	CBORE
Counterclockwise	CCW
Counter-Drill	CDRILL
Countersink	CSK
Cubic	CU
Cylinder	CYL

D

Datum	DAT
Decimal	DEC

Decrease	DECR
Degree	DEG
Detail	DET
Detector	DET
Developed Length	DL
Developed Width	DW
Deviation	DEV
Diagonal	DIAG
Diagram	DIAG
Diameter	DIA
Diameter Bolt Circle	DBC
Diametral Pitch	DP
Dimension	DIM
Direct Current	DC
Disconnect	DISC
Double-Pole Double-Throw	DPDT
Double-Pole Single-Throw	DPST
Dowel	DWL
Draft	DFT
Drafting Room Manual	DRM
Drawing	DWG
Drawing Change Notice	DCN
Drill	DR
Drop Forge	DF
Duplicate	DUP

E

Each	EA
Eccentric	ECC
Effective	EFF
Electric	ELEC
Enclosure	ENCL
Engine	ENG
Engineer	ENGR
Engineering	ENGRG
Engineering Change Order	ECO
Engineering Order	EO
Equal	EQ
Equivalent	EQUIV
Estimate	EST

F

Fabricate	FAB
Fillet	FIL
Finish	FIN
Finish All Over	FAO
Fitting	FTG
Fixed	FXD
Fixture	FIX
Flange	FLG

STANDARD ABBREVIATIONS
(Continued)

Flat Head	FHD	**J**		Modification	MOD
Flat Pattern	F/P			Mold Line	ML
Flexible	FLEX	Joggle	JOG	Motor	MOT
Fluid	FL	Junction	JCT	Mounting	MTG
Forged Steel	FST			Multiple	MULT
Forging	FORG				
Furnish	FURN	**K**			
				N	
		Keyway	KWY		
G				Nickel Steel	NS
				Nomenclature	NOM
Gage	GA	**L**		Nominal	NOM
Gallon	GAL			Normalize	NORM
Galvanized	GALV	Laboratory	LAB	Not to Scale	NTS
Gasket	GSKT	Lacquer	LAQ	Number	NO.
Generator	GEN	Laminate	LAM		
Grind	GRD	Left Hand	LH		
Ground	GRD	Length	LG	**O**	
		Letter	LTR		
		Limited	LTD	Obsolete	OBS
H		Limit Switch	LS	Opposite	OPP
		Linear	LIN	Oscilloscope	SCOPE
Half-Hard	1/2H	Liquid	LIQ	Ounce	OZ
Handle	HDL	List of Material	L/M	Outside Diameter	OD
Harden	HDN	Long	LG	Over-All	OA
Head	HD	Low Carbon	LC		
Heat Treat	HT TR	Low Voltage	LV		
Hexagon	HEX	Lubricate	LUB	**P**	
High Carbon Steel	HCS				
High Frequency	HF			Package	PKG
High Speed	HS	**M**		Parting Line (Castings)	PL
Horizontal	HOR			Parts List	P/L
Hot-Rolled Steel	HRS	Machine(ing)	MACH	Pattern	PATT
Hour	HR	Magnaflux	M	Piece	PC
Housing	HSG	Magnesium	MAG	Pilot	PLT
Hydraulic	HYD	Maintenance	MAINT	Pitch	P
		Major	MAJ	Pitch Circle	PC
		Malleable	MALL	Pitch Diameter	PD
I		Malleable Iron	MI	Plan View	PV
		Manual	MAN	Plastic	PLSTC
Identification	IDENT	Manufacturing (ed, er)	MFG	Plate	PL
Inch	IN	Mark	MK	Pneumatic	PNEU
Inclined	INCL	Master Switch	MS	Port	P
Include, Including,	INCL	Material	MATL	Positive	POS
Inclusive		Maximum	MAX	Potentiometer	POT
Increase	INCR	Measure	MEAS	Pounds Per Square Inch	PSI
Independent	INDEP	Mechanical	MECH	Pounds Per Square	PSIG
Indicator	IND	Medium	MED	Inch Gage	
Information	INFO	Meter	MTR	Power Amplifier	PA
Inside Diameter	ID	Middle	MID	Power Supply	PWR SPLY
Installation	INSTL	Military	MIL	Pressure	PRESS
International Standards	ISO	Millimeter	MM	Primary	PRI
Organization		Minimum	MIN	Process, Procedure	PROC
Interrupt	INTER	Miscellaneous	MISC	Product, Production	PROD

STANDARD ABBREVIATIONS
(Continued)

Q

Quality	QUAL
Quantity	QTY
Quarter-Hard	1/4H

R

Radar	RDR
Radio	RAD
Radio Frequency	RF
Radius	RAD or R
Ream	RM
Receptacle	RECP
Reference	REF
Regular	REG
Regulator	REG
Release	REL
Required	REQD
Resistor	RES
Revision	REV
Revolutions Per Minute	RPM
Right Hand	RH
Rivet	RIV
Rockwell Hardness	RH
Round	RD

S

Schedule	SCH
Schematic	SCHEM
Screw	SCR
Screw Threads	
American National Coarse	NC
American National Fine	NF
American National Extra Fine	NEF
American National 8 Pitch	8N
American Standard Taper Pipe	NTP
American Standard Straight Pipe	NPSC
American Standard Taper (Dryseal)	NPTF
American Standard Straight (Dryseal)	NPSF
Unified Screw Thread Coarse	UNC

Unified Screw Thread Fine	UNF
Unified Screw Thread Extra Fine	UNEF
Unified Screw Thread 8 Thread	8UN
Section	SECT
Sequence	SEQ
Serial	SER
Serrate	SERR
Sheathing	SHTHG
Sheet	SH
Silver Solder	SILS
Single-Pole Double-Throw	SPDT
Single-Pole Single-Throw	SPST
Society of Automotive Engineers	SAE
Solder	SLD
Solenoid	SOL
Speaker	SPKR
Special	SPL
Specification	SPEC
Spot Face	SF
Spring	SPG
Square	SQ
Stainless Steel	SST
Standard	STD
Steel	STL
Stock	STK
Support	SUP
Switch	SW
Symbol	SYM
Symmetrical	SYM
System	SYS

T

Tabulate	TAB
Tangent	TAN
Tapping	TAP
Technical Manual	TM
Teeth	T
Television	TV
Temper	TEM
Temperature	TEM
Tensile Strength	TS
Thick	THK
Thread	THD
Through	THRU

Tolerance	TOL
Tool Steel	TS
Torque	TOR
Total Indicator Reading	TIR
Transformer	XFMR
Transistor	XSTR
Transmitter	XMTR
Tungsten	TU
Typical	TYP

U

Ultra-High Frequency	UHF
Unit	U
Universal	UNIV
Unless Otherwise Specified	UOS

V

Vacuum	VAC
Vacuum Tube	VT
Variable	VAR
Vernier	VER
Vertical	VERT
Very High Frequency	VHF
Vibrate	VIB
Video	VD
Void	VD
Volt	V
Volume	VOL

W

Washer	WASH
Watt	W
Weatherproof	WP
Weight	WT
Wide, Width	W
Wire Wound	WW
Wood	WD
Wrought Iron	WI

Y

Yield Point (PSI)	YP
Yield Strength (PSI)	YS

TWIST DRILL DATA					
METRIC DRILL SIZES (mm)[1]		Decimal Equivalent in Inches (Ref)	METRIC DRILL SIZES (mm)[1]		Decimal Equivalent in Inches (Ref)
Preferred	Available		Preferred	Available	
	.40	.0157	1.70		.0669
	.42	.0165		1.75	.0689
	.45	.0177	1.80		.0709
	.48	.0189		1.85	.0728
.50		.0197	1.90		.0748
	.52	.0205		1.95	.0768
.55		.0217	2.00		.0787
	.58	.0228		2.05	.0807
.60		.0236	2.10		.0827
	.62	.0244		2.15	.0846
.65		.0256	2.20		.0866
	.68	.0268		2.30	.0906
.70		.0276	2.40		.0945
	.72	.0283	2.50		.0984
.75		.0295	2.60		.1024
	.78	.0307		2.70	.1063
.80		.0315	2.80		.1102
	.82	.0323		2.90	.1142
.85		.0335	3.00		.1181
	.88	.0346		3.10	.1220
.90		.0354	3.20		.1260
	.92	.0362		3.30	.1299
.95		.0374	3.40		.1339
	.98	.0386		3.50	.1378
1.00		.0394	3.60		.1417
	1.03	.0406		3.70	.1457
1.05		.0413	3.80		.1496
	1.08	.0425		3.90	.1535
1.10		.0433	4.00		.1575
	1.15	.0453		4.10	.1614
1.20		.0472	4.20		.1654
1.25		.0492		4.40	.1732
1.30		.0512	4.50		.1772
	1.35	.0531		4.60	.1811
1.40		.0551	4.80		.1890
	1.45	.0571	5.00		.1969
1.50		.0591		5.20	.2047
	1.55	.0610	5.30		.2087
1.60		.0630		5.40	.2126
	1.65	.0650	5.60		.2205
				5.80	.2283

1 Metric drill sizes listed in the "Preferred" column are based on the R'40 series of preferred numbers shown in the ISO Standard R497. Those listed in the "Available" column are based on the R80 series from the same document.

TWIST DRILL DATA (CONTINUED)					
METRIC DRILL SIZES (mm)[1]		Decimal Equiv-alent in Inches (Ref)	METRIC DRILL SIZES (mm)[1]		Decimal Equiv-alent in Inches (Ref)
Preferred	Available		Preferred	Available	
6.00		.2362		19.50	.7677
	6.20	.2441	20.00		.7874
6.30		.2480		20.50	.8071
	6.50	.2559	21.00		.8268
6.70		.2638		21.50	.8465
	6.80[2]	.2677	22.00		.8661
	6.90	.2717		23.00	.9055
7.10		.2795	24.00		.9449
	7.30	.2874	25.00		.9843
7.50		.2953	26.00		1.0236
	7.80	.3071		27.00	1.0630
8.00		.3150	28.00		1.1024
	8.20	.3228		29.00	1.1417
8.50		.3346	30.00		1.1811
	8.80	.3465		31.00	1.2205
9.00		.3543	32.00		1.2598
	9.20	.3622		33.00	1.2992
9.50		.3740	34.00		1.3386
	9.80	.3858		35.00	1.3780
10.00		.3937	36.00		1.4173
	10.30	.4055		37.00	1.4567
10.50		.4134	38.00		1.4961
	10.80	.4252		39.00	1.5354
11.00		.4331	40.00		1.5748
	11.50	.4528		41.00	1.6142
12.00		.4724	42.00		1.6535
12.50		.4921		43.50	1.7126
13.00		.5118	45.00		1.7717
	13.50	.5315		46.50	1.8307
14.00		.5512	48.00		1.8898
	14.50	.5709	50.00		1.9685
15.00		.5906		51.50	2.0276
	15.50	.6102	53.00		2.0866
16.00		.6299		54.00	2.1260
	16.50	.6496	56.00		2.2047
17.00		.6693		58.00	2.2835
	17.50	.6890	60.00		2.3622
18.00		.7087			
	18.50	.7283			
19.00		.7480			

1 Metric drill sizes listed in the "Preferred" column are based on the R'40 series of preferred numbers shown in the ISO Standard R497. Those listed in the "Available" column are based on the R80 series from the same document.

2 Recommended only for use as a tap drill size.

Exploring Metric Drafting

Millimetres	X	.0394	=	Inches
Millimetres	=	25.400	X	Inches
Centimetres	X	.394	=	Inches
Centimetres	=	2.54	X	Inches
Metres	X	3.2809	=	Feet
Metres	=	.3048	X	Feet
Kilometres	X	.6214	=	Miles
Kilometres	=	1.6093	X	Miles
Square centimetres	X	.1550	=	Square inches
Square centimetres	=	6.4515	X	Square inches
Square metres	X	10.7641	=	Square feet
Square metres	=	.0929	X	Square feet
Square kilometres	X	247.1098	=	Acres
Square kilometres	=	.0041	X	Acres
Hectares	X	2.471	=	Acres
Hectares	=	.4047	X	Acres
Cubic centimetres	X	.0610	=	Cubic inches
Cubic centimetres	=	16.3866	X	Cubic inches
Cubic metres	X	35.3156	=	Cubic feet
Cubic metres	=	.0283	X	Cubic feet
Cubic metres	X	1.308	=	Cubic yards
Cubic metres	=	.765	X	Cubic yards
Litres	X	61.023	=	Cubic inches
Litres	=	.0164	X	Cubic inches
Litres	X	.2642	=	U. S. Gallons
Litres	=	3.7854	X	U. S. Gallons
Grams	X	15.4324	=	Grains
Grams	=	.0648	X	Grains
Grams	X	.0353	=	Ounces, avoirdupois
Grams	=	28.3495	X	Ounces, avoirdupois
Kilograms	X	2.2046	=	Pounds
Kilograms	=	.4536	X	Pounds
Kilopascals	X	.145	=	Pounds per square inch
Kilopascals	=	.895	X	Pounds per square inch
Newton-metres	X	1.3558	=	Pound feet
Newton-metres	=	.7376	X	Pound feet
Metric tons (1 000 kilograms)	X	1.1023	=	Tons (2000 pounds)
Metric tons	=	.9072	X	Tons (2000 pounds)
Kilowatts	X	1.3405	=	Horsepower
Kilowatts	=	.746	X	Horsepower
Calories	X	3.9683	=	Btu units
Calories	=	.2520	X	Btu units

mm	in.	mm	in.	mm	in.	mm	in.
.01	.0003937	.11	.00433	.21	.00827	.31	.01221
.02	.00079	.12	.00472	.22	.00866	.32	.01259
.03	.00118	.13	.00512	.23	.00905	.33	.01299
.04	.00157	.14	.00551	.24	.00945	.34	.01338
.05	.00197	.15	.00591	.25	.00984	.35	.01378
.06	.00236	.16	.00630	.26	.01024	.40	.01575
.07	.00276	.17	.00669	.27	.01063	.45	.01772
.08	.00315	.18	.00708	.28	.01102	.50	.01968
.09	.00354	.19	.00748	.29	.01141	1.00	.03937
.10	.00394	.20	.00787	.30	.01181		

INDEX